T0213522

Lecture Notes in Artificial Intelligence   9812

Subseries of Lecture Notes in Computer Science

LNAI Series Editors

Randy Goebel
University of Alberta, Edmonton, Canada
Yuzuru Tanaka
Hokkaido University, Sapporo, Japan
Wolfgang Wahlster
DFKI and Saarland University, Saarbrücken, Germany

LNAI Founding Series Editor

Joerg Siekmann
DFKI and Saarland University, Saarbrücken, Germany

More information about this series at http://www.springer.com/series/1244

Andrey Ronzhin · Gerhard Rigoll
Roman Meshcheryakov (Eds.)

Interactive Collaborative Robotics

First International Conference, ICR 2016
Budapest, Hungary, August 24–26, 2016
Proceedings

 Springer

Editors
Andrey Ronzhin
Russian Academy of Sciences
SPIIRAS
St. Petersburg
Russia

Roman Meshcheryakov
TUSUR
Tomsk
Russia

Gerhard Rigoll
Technical University of Munich
Munich
Germany

ISSN 0302-9743   ISSN 1611-3349   (electronic)
Lecture Notes in Artificial Intelligence
ISBN 978-3-319-43954-9   ISBN 978-3-319-43955-6   (eBook)
DOI 10.1007/978-3-319-43955-6

Library of Congress Control Number: 2016946631

LNCS Sublibrary: SL7 – Artificial Intelligence

© Springer International Publishing Switzerland 2016
This work is subject to copyright. All rights are reserved by the Publisher, whether the whole or part of the material is concerned, specifically the rights of translation, reprinting, reuse of illustrations, recitation, broadcasting, reproduction on microfilms or in any other physical way, and transmission or information storage and retrieval, electronic adaptation, computer software, or by similar or dissimilar methodology now known or hereafter developed.
The use of general descriptive names, registered names, trademarks, service marks, etc. in this publication does not imply, even in the absence of a specific statement, that such names are exempt from the relevant protective laws and regulations and therefore free for general use.
The publisher, the authors and the editors are safe to assume that the advice and information in this book are believed to be true and accurate at the date of publication. Neither the publisher nor the authors or the editors give a warranty, express or implied, with respect to the material contained herein or for any errors or omissions that may have been made.

Printed on acid-free paper

This Springer imprint is published by Springer Nature
The registered company is Springer International Publishing AG Switzerland

Preface

The First International Conference on Interactive Collaborative Robotics (ICR) was established as a satellite event of the 18th International Conference on Speech and Computer (SPECOM) by St. Petersburg Institute for Informatics and Automation of the Russian Academy of Science (SPIIRAS, St. Petersburg, Russia), Munich University of Technology (TUM, Munich, Germany), and Tomsk State University of Control Systems and Radioelectronics (TUSUR, Tomsk, Russia).

Challenges of human–robot interaction that are usually discussed during SPECOM have become so real and pressing that they encouraged organizers to start a new conference and invite researchers in the area of social robotics and collaborative robotics to share experiences in human–machine interaction research and development of robotic and cyberphysical systems.

ICR 2016 was hosted by the Budapest University of Technology and Economics (BME) and the Scientific Association for Infocommunications (HTE), in cooperation with SPIIRAS, TUM, and TUSUR. The conference was held during August 24–26, 2016, in the Aquincum Hotel Budapest located in a prime area alongside the river Danube, on the Buda side of this magnificent city and across the river from the serene Margaret Island, with its famous thermal waters.

This volume contains a collection of submitted papers presented at the conference, which were thoroughly reviewed by members of the Program Committee consisting of around 20 top specialists in the conference topic areas. Theoretical and more general contributions were presented in common (plenary) sessions. Problem-oriented sessions as well as panel discussions brought together specialists in limited problem areas with the aim of exchanging knowledge and skills resulting from research projects of all kinds.

We would like to express our gratitude to the authors for providing their papers on time, to the members of the conference reviewing team and Program Committee for their careful reviews and paper selection, and to the editors for their hard work preparing this volume. Special thanks are due to the members of the local Organizing Committee for their tireless effort and enthusiasm during the conference organization.

August 2016

Andrey Ronzhin
Gerhard Rigoll
Roman Meshcheryakov

Organization

The conference ICR 2016 was organized by the Budapest University of Technology and Economics (BME) and the Scientific Association for Infocommunications (HTE), in cooperation with St. Petersburg Institute for Informatics and Automation of the Russian Academy of Science (SPIIRAS, St. Petersburg, Russia), Munich University of Technology (TUM, Munich, Germany), and Tomsk State University of Control Systems and Radioelectronics (TUSUR, Tomsk, Russia). The conference website is located at:
http://specom.nw.ru/icr.

Program Committee

Géza Németh, Hungary
Igor Kalyaev, Russia
Alexey Kashevnik, Russia
Gerhard Kraetzschmar, Germany
Dongheui Lee, Germany
Roman Meshcheryakov,
   Russia (Co-chair)
Vladmir Pavlovkiy, Russia

Viacheslav Pshikhopov, Russia
Gerhard Rigoll, Germany (Co-chair)
Andrey Ronzhin, Russia (Co-chair)
Yulia Sandamirskaya, Switzerland
Jesus Savage, Mexico
Hooman Samani, Taiwan
Evgeny Shandarov, Russia
Lev Stankevich, Russia

Organizing Committee

Géza Németh (Chair)
Mátyás Bartalis
Polina Emeleva
Alexey Karpov
Ekaterina Miroshnikova

Péter Nagy
Alexander Ronzhin
Andrey Ronzhin
Anton Saveliev
Mária Tézsla

Acknowledgments

Special thanks to the reviewers, who devoted their valuable time to review the papers and thus helped maintain the high quality of the conference review process.

Contents

A Control Strategy for a Lower Limb Exoskeleton with a Toe Joint

Sergey Jatsun, Sergei Savin$^{(\boxtimes)}$, and Andrey Yatsun

Southwest State University, 50 let Oktyabrya 94, Kursk 305040, Russia
teormeh@inbox.ru, savin@swsu.ru, ayatsun@yandex.ru

Abstract. In this paper a lower limb exoskeleton with a toe joint is studied. A mathematical model of the exoskeleton is presented, and the equations of motion are given. The exoskeleton is controlled with a feedback controller. The control system attempts to move the center of mass of the exoskeleton along the desired trajectory. To find the joint space trajectory that allows to perform the desired motion a numerical optimization-based iterative algorithm for solving inverse kinematics is given. The algorithm allows to engage and disengage the toe joint, based on how close the mechanism is to a singular position. That gives us an automatic human-like toe joint engagement, that can be controlled though certain parameters, which is discussed in the paper. The results of the numerical simulation of the exoskeleton motion are presented.

Keywords: Lower limb · Exoskeleton · Control system · Active toe joint · Verticalization

1 Introduction

An exoskeleton is a wearable device that enhances the capabilities of the human who uses it. There are applications for exoskeletons in industrial production, warfare, medicine and life style improvement. This includes rehabilitation exoskeletons/Such exoskeletons enable the user to do the tasks they could not do before and also a provide positive influence on the user's health condition [1, 2]. For people who lost the ability to walk these goals can be achieved by the use of a lower limb exoskeleton [3–5].

One of the main challenges in the development of lower limb exoskeletons is the need to design a control system capable of performing a wide range of human-like motions while guaranteeing the safety of the user. There are several ways of generating desired exoskeleton motions, which include methods based on inverse kinematics [6], the use of motion pattern generation [7] and the use of the information about user's state to predict their intended movements [8, 9]. The first two methods are closely related to the approaches adopted in humanoid robotics where significant progress in motion control has been achieved. The main safety issue for lower limb exoskeleton users is the possibility of losing vertical balance and falling. To prevent this from happening specific control methods such as zero-moment point (ZMP) control are adopted [10, 11]. ZMP control provides the criteria that can be used to check if the mechanism is vertically balanced.

© Springer International Publishing Switzerland 2016
A. Ronzhin et al. (Eds.): ICR 2016, LNAI 9812, pp. 1–8, 2016.
DOI: 10.1007/978-3-319-43955-6_1

Although significant progress in lower limb exoskeleton design has been achieved there is still room for improvement, because of the possibility of adopting a more anthropomorphic exoskeleton structure. An example of such improvement is the introduction of a toe joint. Most of the modern exoskeletons have rigid feet, which limits their motion capabilities and makes their movements less natural. In papers [12, 13] it was shown that introducing a toe joint into a humanoid robot design allows the robot to perform a wider range of motions and can lead to better overall performance. The introduction of a toe joint into a rehabilitation exoskeleton can allow the use of more complex therapy procedures.

In this paper we consider a lower limb exoskeleton with a toe joint performing verticalization motion. The objective of the paper is to present an algorithm that allows automatic engagement of the toe joint, while maintaining balance of the mechanism.

2 Mathematical Model of the Exoskeleton

In this paper we study an exoskeleton consisting of two legs and connected to a torso via active rotational joints. Each leg includes four links (thigh, shin, foot and toe) connected in series via a rotational joint equipped with a motor. We consider the case when the toe links remain motionless on the ground at all times during the verticalization process. We assume that the links are connected to the parts of the human body is such a way that the human and exoskeleton joint axes coincide. The motion takes place in a sagittal plane. In papers [6, 14] it was suggested that this type of exoskeleton motion can be accurately modeled by a planar mechanism model. A diagram of the model is shown in Fig. 1.

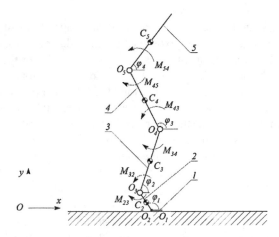

Fig. 1. Diagram of a lower limb exoskeleton with a toe joint; 1 – toe link, 2 – foot link, 3 – shin link, 4 – thing link, 5 – torso link

In Fig. 1 Oxy is the ground reference frame, points $O_2 - O_5$ are active joints, points $C_i(i = \overline{1,5})$ are the centers of mass of the links, $M_{i+1,i}$ are the torques produced by the motors and φ_i are the angles that determine the orientation of the links relative to the horizontal axis Ox. Point O_6 is the end of the 5-th link of the mechanism. The masses of the links are given as the sum of the masses of the links of the exoskeleton and the human body parts they are attached to. The mass distribution of the human body can be found in [15].

For further derivations we introduce a vector of generalized coordinates $\mathbf{q}$:

$$\mathbf{q} = [\varphi_2 \quad \varphi_3 \quad \varphi_4 \quad \varphi_5]^{\mathrm{T}}. \tag{1}$$

It is possible to describe the system with only four generalized coordinates because of the assumption that the toe link remains motionless at all times during the verticalization process. The equations of motion of the system are given in vector form in the following way:

$$\mathbf{A}(\mathbf{q})\,\ddot{\mathbf{q}} + \mathbf{C}(\mathbf{q}, \dot{\mathbf{q}}) + \mathbf{G}(\mathbf{q}) + \boldsymbol{\Phi}(\dot{\mathbf{q}}) = \mathbf{BM}, \tag{2}$$

where $\mathbf{A}(\mathbf{q})$ is a joint space inertia matrix, $\mathbf{C}(\mathbf{q}, \dot{\mathbf{q}})$ is a vector of generalized Coriolis and normal inertial forces, $\mathbf{G}(\mathbf{q})$ is a vector of generalized potential forces, $\boldsymbol{\Phi}(\dot{\mathbf{q}})$ is a vector of generalized dissipative forces, $\mathbf{M}$ is a vector of motor torques, and $\mathbf{B}$ is a linear operator that transforms the vector of motor torques into the vector of the generalized forces. Algorithms for calculating the mentioned vectors and matrices, as well as detailed discussion of their properties can be found in [16].

The given Eq. (5) can be used t model the motion of the system. Some of the expressions (6) and (7) will be used in the controller design in the next chapter.

3   Control System Design

In this section we consider the design of the control system that realizes verticalization motion of the robot. The control system uses a pre-generated desired trajectory of the center of mass of the system, and then uses an inverse kinematics algorithm to derive the desired time functions of generalized coordinates, which are used as an input for a feedback controller. A diagram of the control system is shown in Fig. 2.

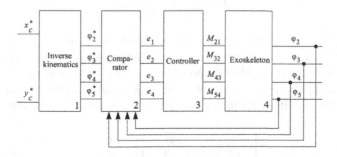

Fig. 2. Diagram of the control system

In Fig. 2 x_C^* and y_C^* are the desired coordinates of the center of mass, φ_i^* are the desired values of the generalized coordinates and e_i^* are components of the control error vector:

$$\mathbf{e} = [\, e_1 \quad e_2 \quad e_3 \quad e_4 \,]^T = \mathbf{q} - \mathbf{q}^*, \tag{3}$$

where $\mathbf{q}^*$ is the vector of the desired values of the generalized coordinates, defined in the same way as $\mathbf{q}$. The values of x_C^* and y_C^* can be found using ZMP control methodology (as it was done in [17]) or directly given by polynomial functions, as it was done in [6, 18]. Here we will consider the later case.

The control actions of the regulator are given by the following equation:

$$\mathbf{M} = \mathbf{B}^{-1}\mathbf{A}(\ddot{\mathbf{q}}^* + \mathbf{K}_p\mathbf{e} + \mathbf{K}_d\dot{\mathbf{e}}), \tag{4}$$

where $\mathbf{K}_p$ and $\mathbf{K}_d$ are diagonal gain matrices with positive elements. The derivation of this controller and the discussion of its properties can be found in [19]. The general theory of such feedback controllers is presented in [20]. The method of tuning the gain matrices $\mathbf{K}_p$ and $\mathbf{K}_d$ is given in paper [21].

A numerical optimization-based algorithm is used to solve the inverse kinematics problem. Such approaches have been adapted in humanoid robotics, where robots such as the Atlas use numerical optimization algorithms to solve the inverse kinematics problem with an onboard computer while the robot is operating [22, 23]. The proposed here algorithm works in two stages. During the first stage it checks whether or not the required position of the body should be obtained without engaging the toe joint. If the toe joint should not be engaged then on the second stage the algorithm finds such orientations of the shin, thigh and torso that the center of mass is placed in the desired position. This is done via optimization over the vector of decision variables $\mathbf{q}$, with the value of ϕ_2 restricted to be equal to π. If the toe joint needs to be engaged than this restriction is being taken off.

The decision on whether or not the toe joint should be engaged is made based upon how close the mechanism is to a singular position. To measure how close the mechanism is to a singular position we introduce the following matrix $\mathbf{J}$:

$$\mathbf{J} = \frac{\partial \mathbf{r}_{C5}}{\partial \mathbf{q}} \left(\frac{\partial \mathbf{r}_{C5}}{\partial \mathbf{q}} \right)^T, \tag{5}$$

where $\mathbf{r}_{C5}$ is the radius vector that describe the position of the center of mass of the torso link of the robot. The matrix $\mathbf{J}$ is a square four by four, and it becomes singular when the mechanism enters a singular position. Its condition number $\kappa(\mathbf{J})$ gets larger as the mechanism approaches a singular position, which allows us to use it as an indicator. The work of the algorithm during the first stage can be described as follows:

$$\begin{cases} \varphi_2 = \pi & \text{if } \kappa(\mathbf{J}) < \kappa_{\max} \\ \alpha_1 \leq \varphi_2 \leq \pi & \text{if } \kappa(\mathbf{J}) \geq \kappa_{\max} \end{cases}, \tag{6}$$

where α_1 is a constant that defines the restriction in the range of motion of the toe joint and κ_{max} is a threshold value for the condition number of $\mathbf{J}$. There are also additional constraints placed on the decision variables. These constraints are there because the human body has restricted ranges of motion in the joints:

$$\alpha_2 \leq \varphi_3 - \varphi_2 \leq \alpha_3, \ \alpha_4 \leq \varphi_4 - \varphi_3 \leq \alpha_5, \ \alpha_6 \leq \varphi_5 - \varphi_4 \leq \alpha_7, \tag{7}$$

where α_i are the constants that determine the range of possible motions in the joints of the exoskeleton user. They can be either individually measured using standard procedures or obtained from the literature [24]. Relations (8) and (9) form the set of constraints for the optimization problem.

On the second stage the proposed algorithm minimizes the following objective function:

$$J_1(\mathbf{q}, t) = \left\| \mathbf{r}_C(\mathbf{q}) - \mathbf{r}_C^*(t) \right\|. \tag{8}$$

It should be noted that the objective function (10) depends on time, which reflects the iterative nature of the algorithm – it needs to be run for every point of time where a solution of the inverse kinematics problem is needed.

The formula for the desired joint space trajectories $\mathbf{q}^*(t)$ obtained by the algorithm has the following form:

$$\mathbf{q}^*(t) = \arg \min_{\mathbf{q}} J_1(\mathbf{q}, t). \tag{9}$$

The resulting desired joint space trajectories $\mathbf{q}^*(t)$ are smoothed by an averaging filter before being used as inputs for the control system.

4 Numerical Simulation

In this section we study the controlled motion of the system. In Fig. 3 the time functions of the generalized coordinates are shown.

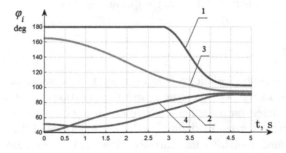

Fig. 3. The time functions of the generalized coordinates; $1 - \varphi_2(t)$, $2 - \varphi_3(t)$, $3 - \varphi_4(t)$, $4 - \varphi_5(t)$

We can observe the graph $\varphi_2(t)$ shown in Fig. 3 Behave similar to a piece-wise polynomial function. For $t < 2.89$ s $\varphi_2(t) = 180°$, and after that it over the next two seconds it monotonically decreases till it reaches the value of $102.6°$. We can show that the time t_c at which the graph $\varphi_2(t)$ starts to decrease depends on the chosen value of κ_{max}. This is illustrated on the Fig. 4.

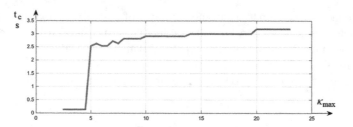

Fig. 4. The dependence of t_c on the value of κ_{max}

The time t_c denotes the moment when the toe joint is being engaged. Analyzing the graph shown in Fig. 4 we can note that for values of κ_{max} less than 4 the engagement of the toe joint happens almost immediately, which means the mechanism performs verticalization while tiptoeing. This may lead to problems with maintaining the vertical balance of the mechanism. When the value of κ_{max} is larger than 5 it shows a more linear relation with t_c. Graph $t_c(\kappa_{max})$ can be used as an instrument for choosing the parameter κ_{max}, such that it would provide the desired toe joint engagement time. It is also possible to demonstrate that the final value of φ_2 is a function of desired value of y_C^* at the end of the motion.

The fact that the presented algorithm only engages the toe joint when the mechanism is close to a singular position mimic the behavior of humans. In many of the human movement the toes start to act only when the leg becomes completely extended or folded (the examples are tiptoe motion, crouching, double support phase of walking). It can be demonstrated that the algorithm can be used to move the exoskeleton to a crouching position where the toe joints will be automatically engaged, because that position is also close to singular. It also should be possible to use the same general principal for toe joint engagement during walking.

5 Conclusions

In this paper a lower limb exoskeleton with a toe joint was considered. A mathematical model of the exoskeleton was presented, and the equations of motion were given. A control system based on a feedback controller was proposed. The inputs for the control system were generated by defining a desired trajectory of the center of mass of the mechanism and solving the inverse kinematics problem. A numerical optimization-based iterative algorithm for solving inverse kinematics was proposed. The algorithm allows to engage and disengage the toe joint, based on how close the mechanism is to a singular position. That gives us an automatic human-like toe joint engagement, that can be controlled though certain parameters that were discussed in the fourth chapter of the paper.

Acknowledgements. Work is supported by RSF, Project № 14-39-00008/.

References

1. Bogue, R.: Exoskeletons and robotic prosthetics: a review of recent developments. Ind. Robot Int. J. **36**(5), 421–427 (2009)
2. Ferrati, F., Bortoletto, R., Menegatti, E., Pagello, E.: Socio-economic impact of medical lower-limb exoskeletons. In: IEEE Advanced Robotics and its Social Impacts (ARSO), pp. 19–26 (2013)
3. Ferris, D.P., Sawicki, G.S., Domingo, A.: Powered lower limb orthoses for gait rehabilitation. Top. Spinal Cord Inj. Rehabil. **11**(2), 34 (2005)
4. Veneman, J.F., Kruidhof, R., Hekman, E.E., Ekkelenkamp, R., Van Asseldonk, E.H., Van Der Kooij, H.: Design and evaluation of the LOPES exoskeleton robot for interactive gait rehabilitation. IEEE Neural Syst. Rehabil. Eng. **15**(3), 379–386 (2007)
5. Jatsun, S., Savin, S., Yatsun, A., Turlapov, R.: Adaptive control system for exoskeleton performing sit-to-stand motion. In: IEEE 10th International Symposium on Mechatronics and its Applications (ISMA), pp. 1–6 (2015)
6. Jatsun, S., Savin, S., Yatsun, A., Malchikov, A.: Study of controlled motion of exoskeleton moving from sitting to standing position. In: Borangiu, T. (ed.) Advances in Robot Design and Intelligent Control. AISC, vol. 371, pp. 165–172. Springer, Heidelberg (2016)
7. Jimenez-Fabian, R., Verlinden, O.: Review of control algorithms for robotic ankle systems in lower-limb orthoses, prostheses, and exoskeletons. Med. Eng. Phys. **34**(4), 397–408 (2012)
8. Dollar, A.M., Herr, H.: Lower extremity exoskeletons and active orthoses: challenges and state-of-the-art. IEEE Trans. Rob. **24**(1), 144–158 (2008)
9. Kawamoto, H., Lee, S., Kanbe, S., Sankai, Y.: Power assist method for HAL-3 using EMG-based feedback controller. In: IEEE International Conference on Systems, Man and Cybernetics, vol. 2, pp. 1648–1653. IEEE (2003)
10. Vukobratovic, M., Hristic, D., Stojiljkovic, Z.: Development of active anthropomorphic exoskeletons. Med. Biol. Eng. **12**(1), 66–80 (1974)
11. Aphiratsakun, N., Parnichkun, M.: Balancing control of AIT leg exoskeleton using ZMP based FLC. Int. J. Adv. Robot. Syst. **6**(4), 319–328 (2009)
12. Nishiwaki, K., Kagami, S., Kuniyoshi, Y., Inaba, M., Inoue, H.: Toe joints that enhance bipedal and fullbody motion of humanoid robots. In: Proceedings of the IEEE International Conference on Robotics and Automation, ICRA 2002, vol. 3, pp. 3105–3110. IEEE (2002)
13. Sellaouti, R., Stasse, O., Kajita, S., Yokoi, K., Kheddar, A.: Faster and smoother walking of humanoid HRP-2 with passive toe joints. In: IEEE/RSJ International Conference on Intelligent Robots and Systems, pp. 4909–4914. IEEE (2006)
14. Jatsun, S., Vorochaeva, L., Yatsun, A., Savin, S.: The modeling of the standing-up process of the anthropomorphic mechanism. In: Proceedings of the 18th International Conference on CLAWAR, Assistive Robotics, p. 175. World Scientific (2015)
15. Plagenhoef, S., Evans, F.G., Abdelnour, T.: Anatomical data for analyzing human motion. Res. Q. Exerc. Sport **54**(2), 169–178 (1983)
16. Featherstone, R.: Rigid Body Dynamics Algorithms. Springer, New York (2014)

17. Panovko, G., Savin, S., Jatsun, S., Yatsun, A.: Simulation of controlled motion of an exoskeleton in verticalization process. J. Mach. Manuf. Reliab. (2016)
18. Jatsun, S.F.: Locomotion control method for patients verticalization with regard to their safety and comfort. In: 26th DAAAM International Symposium on Intelligent Manufacturing and Automation, pp. 1129–1137 (2015)
19. Jatsun S.: Algorithm for motion control of an exoskeleton during verticalization. In: ITM Web of Conferences, vol. 6 (2016)
20. Ortega, R., Spong, M.W.: Adaptive motion control of rigid robots: a tutorial. Automatica **25** (6), 877–888 (1989)
21. Jatsun, S., Savin, S., Yatsun, A.: Parameter optimization for exoskeleton control system using Sobol sequences. In: Proceedings of 21st CISM-IFToMM Symposium on Robot Design (2016)
22. Feng, S., Whitman, E., Xinjilefu, X., Atkeson, C.G.: Optimization based full body control for the atlas robot. In: 14th IEEE-RAS International Conference on Humanoid Robots (Humanoids), pp. 120–127. IEEE (2014)
23. Feng, S., Whitman, E., Xinjilefu, X., Atkeson, C.G.: Optimization-based full body control for the DARPA robotics challenge. J. Field Robot. **32**(2), 293–312 (2015)
24. Roaas, A., Andersson, G.B.: Normal range of motion of the hip, knee and ankle joints in male subjects, 30–40 years of age. Acta Orthop. Scand. **53**(2), 205–208 (1982)

A Recovery Method for the Robotic Decentralized Control System with Performance Redundancy

Iakov Korovin[1], Eduard Melnik[2], and Anna Klimenko[3(✉)]

[1] Southern Federal University, Rostov-on-Don, Russia
[2] Southern Scientific Center of the Russian Academy of Sciences (SSC RAS),
Rostov-on-Don, Russia
[3] Scientific Research Institute of Multiprocessor Computing Systems,
Southern Federal University, Taganrog, Russia
anna_klimenko@mail.ru

Abstract. The fault of the robotic control system is critical and leads to the general system failure, while autonomous robots have to gain their aims without any maintenance. Contemporary academic studies propose decentralized control systems as prospective from the robustness point of view. On the other hand, a performance redundancy allows to optimize resource utilization and improve the fault-tolerance potential of the control system. This paper is devoted to the recovery method of the robotic decentralized control system with performance redundancy. A reconfiguration problem has been formalized, decentralized method of the solution obtaining is represented. Also some simulation results are given and discussed.

Keywords: Decentralized control system · Robustness · Fault-tolerance · Simulated annealing · Autonomous robots control

1 Introduction

Fault-tolerance is extremely important for autonomous robotic systems. The large amount of them is performing their tasks in hazardous and aggressive environments, where a man is not supposed to be located. Besides, autonomous robots perform and must achieve their goals without any repair for a long terms of time.

In robotics, some classifiers of failures are proposed. For example, in [1] the general failure levels are concerned: mechanical level (a joint becomes lock); hardware level (sensor does not perform properly); controller level; controlling computer level.

Another classification of failures is described in [2]: sensors; effectors; communications; power system; control system.

Human and physical faults as a cause of failure are distinguished in [3], where the detailed taxonomy (Fig. 1) also can be found.

In the scope of this paper the robot control systems are under consideration. Contemporary control systems are the software and hardware complexes, where distributed computing paradigm is used widely. The monitoring and control tasks (MCTs)

© Springer International Publishing Switzerland 2016
A. Ronzhin et al. (Eds.): ICR 2016, LNAI 9812, pp. 9–17, 2016.
DOI: 10.1007/978-3-319-43955-6_2

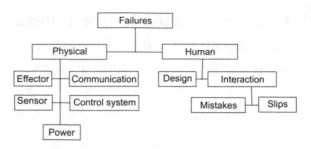

Fig. 1. Failures taxonomy

within the control system are performed by computational units (CUs). Each MCT allocates some computational resources, and each CU performs more then one MCT. The control system architecture from the management point of view can be centralized, hierarchical or decentralized. Some works show that the decentralized control system is potentially more fault-tolerant then others [4], but needs additional research in the fields of cooperative problem solving, multiagent systems, etc.

Author of [5] defines a fault-tolerant control system as a control system, which is able to automatically maintain the system stability and an acceptable performance when component failures occur. To gain these objectives, following principles must be implemented: fault detection; fault isolation; fault identification; fault recovery.

Usually, passive or active recovery methods are used [2]. Within the active recovery methods two main reconfiguration strategies are used. The first one propose using of some pre-defined control laws, the second one is oriented to the on-line synthesis of the system controller with respect to the fault identification.

In the scope of this paper recovery method for the decentralized control system with the performance redundancy will be considered. The next section contains the brief explanation of the performance redundancy in comparison to the structural one. Section 3 is devoted to the reconfiguration problem formalization with the graceful system degradation objective. Section 4 contains reconfiguration method, and, at last, Sect. 5 presents some experimental results and discussion.

2   Performance Redundancy and Decentralized Dispatching

Structural redundancy is widely used nowadays [4]. As mentioned in [5], redundancy is the key ingredient in any fault-tolerant systems. Almost all of modern aircraft such as Boeing 777 and Airbus A320/330/340 have used triplex- or quadriplex-redundant activation systems, flight control computer and databus systems [6, 7].

Performance redundancy considers all CUs as performing elements with some performance reserve. Advantages of performance redundancy are explained in details in [8–10].

The way of ICS dispatching is important too: the centralized dispatching has multiple drawbacks, in particular, the main dispatcher fault is the cause of the entire system failure without the possibility to recover.

Decentralized dispatching of the ICS operates with equal control elements. Each CU is controlled by its own software agent (Fig. 2), which operates as a kind of MCT. In the case of CU failure (software or hardware) agents of operational nodes begin a recovery procedure via reconfiguration: MCTs from the faulted node can be launched by the operational ones.

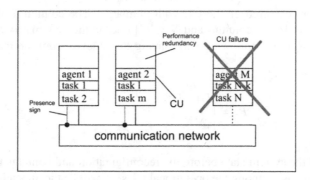

Fig. 2. Performance redundancy and software agents representing the CUs

ICS with decentralized dispatching and performance redundancy has good recover possibilities, but requires the design and implementation of the cooperative recovery methods and algorithms.

3 Reconfiguration Problem Formalization

Let the input data be the following:

- A set of MCTs $G = \{x_i\}$, $i = 1...N$, where x_i – the size of task i, N – the number of tasks.
- Let $G = G_c \cup G_{nc}$, $G_c \cap G_{nc} = \emptyset$, where G_c – a subset of critical MCTs, G_{nc} — a subset of non-critical ones. Non-critical MCTs can be eliminated from the system during reconfiguration. The number of critical MCTs is N_c, and the number of non-critical MCTs N_{nc};
- Let G_f be the set of MCTs from the faulted CU. $G_f \subseteq G$, G_p is the performing tasks, $G_p \subseteq G$, $G_p \cap G_f = \emptyset$.
- A planned completion time for the set G is T_{plan}.
- Number of CUs is M with the performance p.

Let' s take into consideration that we have to allocate the MCTs from the set G_f within the system of operational CUs, on which the tasks from the set G_p are allocated with the constraint of completion time T_{plan}. Let the resource allocated by CU j for the subtask i be λ_{ij}. The tasks allocation before the failure is described by matrix R:

$$R = \begin{vmatrix} r_{11} & r_{12} & r_{1M} \\ \cdots & \cdots & \cdots \\ r_{N1} & \cdots & r_{NM} \end{vmatrix}, \tag{1}$$

where $r_{ij} = f(\frac{x_i}{\lambda_{ij}p})$, $f(\frac{x_i}{\lambda_{ij}p}) = \begin{cases} \frac{x_i}{\lambda_{ij}p}, & \text{if } x_i \text{ is running on CU } j, \\ 0, & \text{otherwise}. \end{cases}$

Let the failure occurred on the CU with number d. The column d of matrix R is deleted, so there is $M - 1$ columns and $N - |G_f|$ lines in the new matrix R_f. Renumber the elements of R_f in the following way, saving the indexes from the matrix R in the upper positions:

$$R_f = \begin{vmatrix} r_{11}^{ij} & r_{12}^{ij} & r_{1(M-1)}^{iM} \\ \cdots & \cdots & \cdots \\ r_{11}^{Nj} & \cdots & r_{(N-G_f)(M-1)}^{NM} \end{vmatrix}. \tag{2}$$

R_f describes the system state before the reconfiguration and contains the allocation of the operational tasks among the operational CUs. R_r will be the allocation of the task set G on the $M - 1$ CUs. Formally, the subset G_f will be added to the G_p with the number of $CU_s = M - 1$:

$$R_r = \begin{vmatrix} r_{11}^{ij} & r_{12}^{ij} & r_{1(M-1)}^{iM} \\ \cdots & \cdots & \cdots \\ r_{11}^{Nj} & \cdots & r_{(N)(M-1)}^{NM} \end{vmatrix}; \tag{3}$$

$$r_{ij}^{kl} = f(\frac{x_i}{\lambda_{ij}p}) g(x_i);$$

$$g(x_i) = \begin{cases} 0, \text{if } x_i \in G_{nc} \text{ and eliminated from the system}, \\ 1, \text{otherwise}. \end{cases} \tag{4}$$

Let's consider matrix Ψ:

$$\Psi = \begin{bmatrix} \varphi(x_{1j}^{kl}) & \varphi(x_{2j}^{kl}) & \cdots & \varphi(x_{Nj}^{kl}) \end{bmatrix} \tag{5}$$

where $\varphi(x_{1j}^{kl}) = \begin{cases} 0, l = j, \\ \xi, \text{otherwise} \end{cases}$ k, l — the saved indexes of matrix R, j — the number of CU in matrix R_r, ξ — the integer number.

The matrix Ψ describes if the MCT x_i was relocated from CU l to CU j.

The first objective function can be written in the following manner:

$$F_1 = \sum_{i=1}^{N} \varphi(x_{ij}^{kl}) \to MIN. \tag{6}$$

The next objective function component is load balancing which can be written in the following way.

$$F_2 = (\sum_{i=1}^{N} \lambda_{ik} - \sum_{i=1}^{N} \lambda_{il}) \rightarrow MIN, \forall k, l, \lambda_{ik} \subset R_r. \tag{7}$$

In other words, we need to find MCTs allocation with respect to load balancing objective function. While the desirable option is to deliver the graceful system degradation, it is useful to keep running as much MCTs as possible. The maximum number of non-critical tasks running equals to the maximum summa of all $g(x_i)$ in the matrix R'.

The last objective function component will be as following:

$$F_3 = -\sum_{i}^{N} g(x_i) \rightarrow MIN. \tag{8}$$

Herewith the time constraint must be satisfied:

$$\forall j : \sum_{i=1}^{N} r'_{ij} \leq T_{plan}, j \in [1...M]. \tag{9}$$

Let's put the current multicriteria optimization problem to the following form:

$$F = \sum_{i=1}^{N} \varphi(x_{ij}^{kl}) \rightarrow MIN, \tag{10}$$

$$(\sum_{i=1}^{N} \lambda_{ik} - \sum_{i=1}^{N} \lambda_{il}) \leq \gamma; \ -\sum_{i}^{N} g(x_i) \leq \mu; \forall j : \sum_{i=1}^{N} r'_{ij} \leq T_{plan}, j \in [1...M],$$

$x_i > 0; \ 0 < \lambda_{ij} < 1$, where $0 < \gamma < 1$ is the assumed level of load dispersing, μ is the integer number.

4   Cooperative Problem Solving

The problem formalized earlier contains a kind of k-partition problem (or bean-packing problem) which is NP-hard, so there is no polynomial algorithms for the solution obtaining. In the scope of this research the simulated annealing (SA) with the "quenching" temperature is used [11] to reach an acceptable solution in a reasonable time.

With the shortage of time and the using of decentralized control, it is appropriate to initiate a search for a new system configuration at all operational nodes (which are represented by the agents, Fig. 2).

After CU or MCT fault detection and identification (which are out of this paper's scope), the reconfiguration is initialized, and every performing agents launches the new configuration search.

As soon as one of the agents finds allowable configuration, solving, in fact, a constraint satisfaction problem, it notifies other agents, which take the solution found as a new configuration proposed to perform.

Here we have to make some assumptions. The system of agent is synchronous in terms of work [12]. If one agent sends a message, agent-addressee receives it without delay. There is no message losses in the communication network. The model of communication network is fully connected graph. Then, the next assumption takes place: some agents can broadcast incorrect solution as a result of search. So, the cooperative configuration search method must have a kind of mechanism to prevent the further usage of unviable solution.

Each agent also must have a queue (Q) for the incoming messages and a queue (S) for the viable solution. Besides, we assume every agent know the constraints for the MCTs: launching time spans, data transfer interconnections, etc.

Generalized method based on a simulated annealing for one agent is represented below:

1. Set the initial parameters: temperature, quenching ratio, etc.
2. Generate solution R in a random manner.
3. If Q contains any solutions, go to the 4.
4. Beginning of the cycle
4.1. If Q contains any solutions, go to the 5:
4.2. Generation of new solutions: R.
4.3. Calculate the value of F.
4.4. Check the admissibility of F
4.5. If the current solution is acceptable, go to 5.
4.6. Temperature correction. Go to step 4.
5. Broadcast the solution reached.
5.1. Range the solutions in the Q.
5.2. Verify the best solution. If the verification is successful, go to 6.
5.3. If the solution is unviable, delete it from Q. Go to 5.2.
6. Broadcast the verified solution.
7. Choose the most frequent viable solution from S for the execution.
8. End.

The cooperative method described above allows every agent to have all solutions in Q after at least one agent found a solution. Verification process contains the check of constraints for the MCTs. The S queue contains solutions estimated as "viable". We assume that if solution S_1 was accepted by N_1 agents, and solution S_2 was accepted by N_2 agents, S_1 is "viable" if $N_1/N_2 = 2$ [12].

5 Simulation Results and Brief Discussion

Taking into consideration the shortage of time, the first simulation study is in the field of solution obtaining speed. For this study SA with Boltzmann generation rule and quenching ratios 0.9; 0.8; 0.7 was used (Fig. 3).

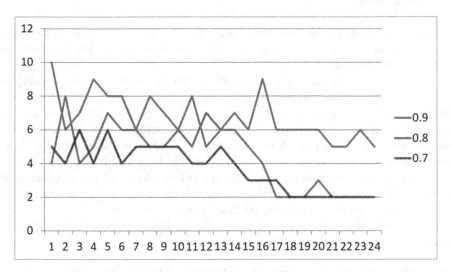

Fig. 3. SA with "quenching" temperature scheme convergence speed

The results of pilot simulation allow to affirm that some local minimas can be reached fast enough (10–20 iterations). It makes SA a perspective search method even in the time shortage circumstances (2 eliminated non-critical tasks as a result).

Next simulation is made for the different number of calculating agents (5;10) with the initial number of MCTs = 50 (Fig. 4).

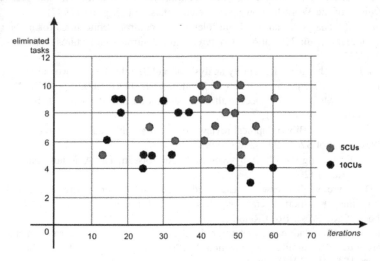

Fig. 4. The number of eliminated MCTs within 5 and 10CUs cooperative problem solving

It is seen, that with the increasing of agents number the quality and the speed of the solution obtaining becomes better. The best result for the 5 agents is 5 eliminated non-critical tasks, while the best result for the 10 agents is 3 lost tasks.

6   Conclusions

Fault-tolerance is one of the important aspects for the autonomous robots. An enormous field of tasks is performed without man's assistance, so the robots should be reliable, viable and fault-tolerant.

In the scope of this paper the decentralized robotic control system recovery is considered. A reconfiguration problem is formalized, and a method for the cooperative problem solving is proposed. A method represented bases on the parallel multistart SA and provides some degree of robustness in the circumstances of incorrect agent behavior. It must be noted, that SA with "quenching" temperature scheme can be "fast" enough to find local minimas. Also, a group of searching agents improve solution quality significantly. For example, if one agent can find a local minima of unaccepted quality, another one, with different initial computational point, can find the acceptable solution. It must be noted, that the redundancy must be sufficient.

The future work is proposed to be directed to the field of robust distributed algorithms and one's modelling and efficiency estimation.

Acknowledgements. The reported study was funded by SSC RAS project 0256-2014-0008 within the task 007-01114-16 PR and by RFBR projects 14-08-00776-a and 15-37-20821-mol-a-ved.

References

1. Gini, M., Smith, R.: Monitoring robot actions for error detection and recovery. In: Proceedings of the Workshop on Space Telerobotics, vol. 3, p. 67 (1987)
2. Crestani, D., Godary-Dejean, K.: Fault tolerance in control architectures for mobile robots: fantasy or reality?. In: 7th National Conference on Control Architectures of Robots, Nancy, France (2012)
3. Carlson, J., Murphy, R.R.: Reliability analysis of mobile robots. In: Proceedings of the IEEE International Conference on Robotics and Automation (ICRA 2003), pp. 274–281 (2003)
4. Kalyaev, I.A., Melnik, E.V.: Decentralized Systems of Computer Control. Publishing SSC RAS, Rostov-on-Don (2011)
5. Zhang, Y., Jiang, J.: Bibliographical review on reconfigurable fault-tolerant control systems. Ann. Rev. Control **32**(2), 229–252 (2008)
6. Bartley, G.F.: Boeing B-777: fly-by-wire flight controls. In: The Avionics Handbook. CRC Press, Boca Raton (2001)
7. Biere, D., Favre, C., Traverse, P.: Electrical flight controls, from Airbus A320/330/340 to future military transport aircraft: a family of fault-tolerant systems. In: The Avionic Handbook. CRC Press, Boca Raton (2001)
8. Melnik, E.V.: Simulation options for redundancy in distributed information and control systems with a decentralized organization. In: Proceedings of SFU, SER Technical science, no. 3, pp. 184–193 (2013). (In Russian)

9. Melnik, E.V.: Principles of organization of the decentralized network-centric information management systems. Herald Comput. Inf. Technol. **4**, 25–30 (2013). (In Russian)
10. Melnik, E.V.: Effect processor computational load balancing devices in highly distributed information management system. In: Mechatronics, Automation, Control, pp. 29–35 (2012). (In Russian)
11. Ingber, L.: Simulated annealing: practice versus theory (1993). http://citeseer.uark.edu:8080/citeseerx/viewdoc/summary?doi=10.1.1.15.1046
12. Tel, G.: Introduction to Distributed Algorithms, pp. 1–608. Cambridge University Press, Cambridge (2000)

An Analysis of Visual Faces Datasets

Ivan Gruber[1,2,3]([⊠]), Miroslav Hlaváč[1,3], Marek Hrúz[2], Miloš Železný[1],
and Alexey Karpov[3,4]

[1] Faculty of Applied Sciences, Department of Cybernetics,
UWB, Pilsen, Czech Republic
{mhlavac,zelezny}@kky.zcu.cz
[2] Faculty of Applied Sciences, NTIS, UWB, Pilsen, Czech Republic
{grubiv,mhruz}@ntis.zcu.cz
[3] ITMO University, St. Petersburg, Russia
karpov@iias.spb.su
[4] SPIIRAS, St. Petersburg, Russia

Abstract. This paper presents an analysis of datasets of images of
human faces with annotated facial keypoints, which are important in
human-machine interaction, and their comparison. Datasets are divided
according to external conditions of the subject into two groups: datasets
in laboratory conditions and in the wild data. Moreover, a quick review
of the state-of-the-art methods for keypoints detection is provided. Existing methods are categorized into the following three groups according to
the approach to the solution of the problem: top-down, bottom-up and
their combination.

Keywords: Facial keypoint · Keypoints detection · Facial landmark
localization · Dataset · Computer vision

1 Introduction

The purpose of facial keypoints detection (FKD) is to detect keypoints, also known
as feature points or fiducial points, on a human face, see Fig. 1. This task is very
complex and demanding due to various external conditions, for example, illumination, pose or occlusion, or internal conditions, for example, face expression or
aging.

Despite all these problems, FKD is a very popular task with high relevance in
computer vision with applications in robotics mainly for human-machine interaction. Existing methods can be categorized into three groups. Methods in the first
group use a top-down approach to solve FKD problem. The top-down approach
is essentially every approach, that starts with the big picture of the problem and
breaks it down into smaller segments. In the second group, there are methods,
that use a bottom-up approach. In this approach the problem is first specified in the detail. These details are then linked together to form bigger subsystems, which are also linked together until a complete top-level system is formed.

© Springer International Publishing Switzerland 2016
A. Ronzhin et al. (Eds.): ICR 2016, LNAI 9812, pp. 18–26, 2016.
DOI: 10.1007/978-3-319-43955-6_3

The third approach is a combination of the previous approaches. We provide more information about these methods in Sect. 2.

To create the vast majority of these methods it was required to have data for training and testing. A wide range of datasets (also known as databases) was created for this task around the world. Facial keypoint detection datasets can be divided into two groups according to external conditions of the subject. In the first group of datasets the subject is captured in laboratory conditions, i.e. usually with constant illumination, with a known pose, and without any occlusion. In the second group, there are datasets with images captured in the wild (i.e. with variable external conditions). Detailed information about these datasets can be found in Sect. 3. Some of these datasets, especially in the wild datasets, are used for benchmark testings of new algorithms.

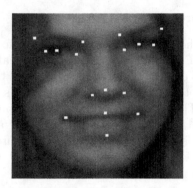

Fig. 1. Facial keypoints (white dots), picture from MUCT database [1].

2   State-of-the-Art Methods

There is a lot of existing algorithms to solve FKD problem nowadays and we provide a quick review of them in this section. The FKD usually starts from a bounding box around detected face returned from a face detector (for example Viola-Jones detector [2]), but some existing algorithms merge these tasks (face detection and FKD) into one. Following steps are different from method to method, however, we can find some common elements and according to these elements we can divide existing methods into some groups. There are many possible divisions of existing methods, we chose to divide them into the following three groups according to the approach to the solution of the problems: top-down approach, bottom-up approach, and their combination.

2.1   Top-Down Approach

The top-down approach starts from the big picture and proceeds down to the smaller segments. The strength of this approach lies in creating highly adapted algorithm for the given problem. A typical example of a method falling into this category is Active Appearance Model (AAM) [3], for example Antakos et al. [4] based their novel generative deformable model on AAM and pictorial structures. Yu et al. [5] successfully employed two-stage cascade deformable shape model to effectively localize facial landmarks from a single camera. In the same year the Robust Cascade Pose Regression method [6] was proposed. In 2014 Asthana et al. proposed Chehra tracker [7] with very good results. The tracker uses a discriminative model that is trained as a cascade of regressors. Moreover, the authors have developed an efficient strategy to update the model. Tzimiropoulos [8] proposed a similar approach, however, he used regression to learn a sequence of average Jacobian and Hessian matrices from the data and from the descent direction. He uses this regression to fit the facial deformable model.

2.2   Bottom-Up Approach

The bottom-up approaches are based on obtaining information directly from images of human faces. FKD then uses RGB or grayscale values of individual pixels or larger regions to determine the exact position of the landmarks. Typical representatives of the bottom-up approach are methods based on deep neural networks (DNN) [9,10]. The whole image serves as an input to the network. This approach is more general but at the cost of complexity and computing power. Other bottom-up methods use regressors to detect feature points [11,12].

2.3   Combination

Because of the specific strengths of top-down and bottom-up approaches some form of combination is desired to obtain methods utilizing the strengths of both. Segmentation-aware Part Model uses graph cut as an extension to general part model to identify occluders [13]. In [14] regression of local binary features was used in combination with ensemble trees.

3   Datasets

In this section, we provide an analysis of the selected datasets. Due to the restrictions of the article range and due to the amount of different datasets, we could not include them all, but we selected the most important ones and those which significantly differs from other datasets. We provide comparisons between datasets in Table 1 and example images (see Fig. 2) at the end of this section.

3.1   LFPW

Labeled Face Parts in the Wild (LFPW) [15] dataset is one of the most popular datasets for keypoints detection benchmarks. The goal of the dataset's creators was to create a set with more challenging conditions (especially pose) than BioID dataset [16].

Table 1. Comparison of datasets from Sect. 3.

Dataset	Number of images	Number of keypoints	Conditions	Resolution
LFPW [15]	1432	35	Variable	Variable
MUCT [1]	3755	76	Laboratory	640 × 480
HELEN [17]	2330	194	Variable	Variable
AFW [18]	205	6	Variable	Variable
AFLW [19]	25993	21	Variable	Variable
COFW [20]	1852	29	Variable	Variable
300-W [21]	4102	68	Variable	Variable
Multi-PIE [23]	75000	39–68	Laboratory	3072 × 2048
XM2VTSDB [24]	2360	68	Laboratory	720 × 576

Fig. 2. Exemplary pictures from different datasets. Starting from top left corner row-by-row it is LFPW [15], MUCT [1], COFW [20], HELEN [17], AFLW [19].

Release 1 of LFPW originally contained 1432 images divided into 1132 training images and 300 test images. The dataset is designed to test facial points detection algorithms in unconstrained conditions. The main advantage of this set is a large variation of faces in different illuminations, poses, and expressions.

However, it shares only a list of image URLs due to copyright issues. That means, that some links are no longer valid. Images were downloaded from the web and were analyzed by an off-the-shelf face detection system. Each of 35 fiducial points was labeled manually by Amazon Mechanical Turk workers.

3.2   MUCT

MUCT database [1] was created to provide more diversity of illumination, age, and ethnicity than landmarked 2D face databases at the time.

The database consists of 3755 RGB images captured in laboratory conditions with 640 × 480 resolution. There are 276 persons of different ages, ethnicities, and both sexes included in the database. Each person was captured from five cameras simultaneously (each camera captured the subject from different angle). More-over, each subject was captured in two or three lighting conditions (ten different lighting setups were used in total). There were 76 keypoints manually annotated on each image. If the keypoint was occluded, it was assigned to coordinates [0,0] (top left corner of the image). Persons have a natural expression in the images.

3.3   HELEN

Vuong et al. [17] effort was to create a dataset with high resolution images with broad range of appearance variation, including pose, illumination, expression, and occlusion. The dataset consists of images from Flickr search. A face detector was run on these images to filter out the ones with a small face area (smaller than 500 pixels in width), then the remaining images were filtered further by hand.

The dataset consists of 2330 high resolution images in variable conditions and with variable resolutions divided into a training (2000 images) and a testing set (330 images). The images were manually annotated using Amazon Mechanical Turk. Each of them contains 194 keypoints on 7 contours.

3.4   AFW

Annotated Faces in the Wild (AFW) dataset [18] contains 205 images with 468 faces with a large variation in appearance (pose, age, ethnicity) and external conditions. Each face is labeled with 6 keypoints (tip of the nose, center of the eyes, corners of the mouth, and the center of the mouth), a bounding box and a discretized viewpoint. The main difference between this dataset and the other "in the wild" sets is in its annotation of multiple non-frontal faces in a single image.

3.5   AFLW

The main motivation to create Annotated Facial Landmarks in the Wild (AFLW) database [19] was the need for a large-scale, multi-view, real-world face database with annotated facial features. Images were gathered on Flickr and then manually searched for images containing faces.

The database consists of 25993 faces (59 % females and 41 % males) with 21 manually annotated facial keypoints (no annotation is present if a facial keypoint is not visible). Most of the images are colored. The main advantage of this dataset is its big variation in pose, external conditions, ethnicity and age of the subjects. Moreover, the database also provides face rectangles and ellipses.

3.6 COFW

Caltech Occluded Faces in the Wild [20] contains 1007 face images in real-world conditions. Images show different levels of occlusion and they are annotated by hand with 29 keypoints. Both occluded and unoccluded feature points are annotated. Average occlusion is 23 %. Dataset is divided into grayscale and color images.

3.7 300-W

This dataset [21] has been created for the 300 Faces-In-The-Wild Challenge 2013. It contains images from LFPW, HELEN, AFW, XM2VTS [22], FRGC ver.2 and IBUG. All images have been re-annotated with unified 68 and 51 keypoints. Images from IBUG dataset contain 135 posses and expressions. Testing data are composed from newly collected 300 outdoor and 300 indoor images.

3.8 Multi-PIE

Multi-PIE [23] is a large dataset created at Carnegie Mellon University in 2010. It contains 337 different subjects, captured from 15 view points with 19 different illuminations. The total number of images is more than 750.000 but only 6152 of them are annotated with AAM-based style labels. The labels have between 39 and 68 keypoints depending on the pose. All points were annotated manually. Images are stored as JPG (high-res) or PNG. The dataset can be ordered on a dedicated USB-attached hard drive with world-wide delivery.

3.9 XM2VTSDB

XM2VTSDB [24] contains 2360 images of 295 subjects annotated by hand with 68 keypoints. Images are in color with a uniform resolution 720×576. Video sequences and corresponding audio recordings are part of the dataset. There are other datasets containing both audio and video used for audiovisual speech recognition (for example UWB-05-HSCAVC [25] or REPERE [26])

4   Conclusion and Future Work

The task of facial keypoints detection is a challenging problem in the field of computer vision that is receiving big attention over the last years. The facial keypoints detection datasets are an integral part of this problem. There are many different datasets available, which can be used in the effort of creating a robust algorithm that can reliably handle broad ranges of lighting conditions, poses, occlusions, expressions or even aging. We provided a brief analysis of these datasets and their comparison. We also provided a short summary of the state-of-the-art facial feature keypoints detection methods. It is hard to orientate in such a fast developing field as facial keypoints detection and it is necessary to spend enormous amount of time to gather all of the information needed before the actual research work. We hope this article will help others with that and simplify their work.

We would like to focus our future research on developing a robust facial keypoints detection method based on deep neural networks. We will utilize some of the datasets as training and testing data. Moreover, we are planning to use the results from the detection for face recognition and lip tracking.

Acknowledgments. This work is supported by grant of the University of West Bohemia, project No. SGS-2016-039, by Ministry of Education, Youth and Sports of Czech Republic, project No. LO1506, by Russian Foundation for Basic Research, project No. 15-07-04415, and by the Government of Russian, grant No. 074-U01.

References

1. Milborrow, S., Morkel, J., Nicolls, F.: The MUCT landmarked face database. Pattern Recogn. Assoc. S. Afr. **201**(0), 179–184 (2010)
2. Viola, P., Jones, M.: Robust real-time face detection. Int. J. Comput. Vis. **57**(2), 137–157 (2004)
3. Cootes, T.F., Taylor, C.J.: Statistical models of appearance for computer vision. In: Imaging Science and Biomedical Engineering, University of Manchester, pp. 149–163 (2004)
4. Antonakos, E., Alabort-i-Medina, J., Zaferiou, S.: Active pictorial structures. In: IEEE Conference on Computer Vision and Pattern Recognition (CVPR), Boston, pp. 5435–5444 (2015)
5. Yu, X., Huang, J., Zhang, S.: Pose-free facial landmark fitting via optimized part mixtures and cascaded deformable shape model. In: IEEE International Conference on Computer Vision, Sydney, pp. 1944–1951 (2013)
6. Burgos-Artizzu, X.P., Perona, P., Dollár, P.: Robust face landmark estimation under occlusion. In: IEEE International Conference on Computer Vision, Sydney, pp. 1513–1520 (2013)
7. Asthana, A., Zaferiou, S., Cheng, S.: Incremental face alignment in the wild. In: IEEE Conference on Computer Vision and Pattern Recognition, Columbus, pp. 3659–3667 (2014)
8. Tzimiropoulos, G.: Project-out cascaded regression with an application to face alignment. In: IEEE Conference on Computer Vision and Pattern Recognition (CVPR), Boston, pp. 3659–3667. IEEE (2015)

9. Zhang, J., Kan, M., Shan, S., Chen, X.: Leveraging datasets with varying annotations for face alignment via deep regression network. In: Proceedings of IEEE International Conference on Computer Vision (ICCV), pp. 3801–3809 (2015)
10. Zhang, Z., Luo, P., Loy, C.C., Tang, X.: Facial landmark detection by deep multi-task learning. In: Proceedings of European Conference on Computer Vision (ECCV), pp. 94–108 (2014)
11. Kazemi, V., Sullivan, J.: One millisecond face alignment with an ensemble of regression trees. In: Proceedings of IEEE Conference on Computer Vision and Pattern Recognition (CVPR), pp. 1867–1874 (2014)
12. Yu, X., Lin, Z., Brandt, J., Metaxas, D.N.: Consensus of regression for occlusion-robust facial feature localization. In: Fleet, D., Pajdla, T., Schiele, B., Tuytelaars, T. (eds.) ECCV 2014, Part IV. LNCS, vol. 8692, pp. 105–118. Springer, Heidelberg (2014)
13. Ghiasi, G., Fowlkes, C.: Using segmentation to predict the absence of occluded parts. In: Proceeding of British Machine Vision Conference (BMVC) (2015)
14. Ren, S., Cao, X., Wei, Y., Sun, J.: Face alignment via regressing local binary features. Proc. IEEE Trans. Image Process. 25(3), 1233–1245 (2016)
15. Belhumeur, P.N., Jacobs, D.W., Kriegman, D.J., Kumar, N.: Localizing parts of faces using a consensus of exemplars. In: Proceedings of 24th IEEE Conference on Computer Vision and Pattern Recognition, vol. 35, no. 12, pp. 2930–2940 (2011)
16. Jesorsky, O., Kirchberg, K.J., Frischholz, R.W.: Robust face detection using the hausdorff distance. In: Bigun, J., Smeraldi, F. (eds.) AVBPA 2001. LNCS, vol. 2091, pp. 90–95. Springer, Heidelberg (2001)
17. Le, V., Brandt, J., Lin, Z., Bourdev, L., Huang, T.S.: Interactive facial feature localization. In: Fitzgibbon, A., Lazebnik, S., Perona, P., Sato, Y., Schmid, C. (eds.) Computer Vision – ECCV 2012. LNCS, vol. 7574, pp. 679–692. Springer, Heidelberg (2012)
18. Zhu, X., Ramanan, D.: Face detection, pose estimation and landmark localization in the wild. In: Computer Vision and Pattern Recognition (CVPR) Providence, Rhode Island (2012)
19. Koestinger, M., Wohlhart, P., Roth, P.M., Bischof, H.: Annotated facial landmarks in the wild: a large-scale, real-world database for facial landmark localization. In: First IEEE International Workshop on Benchmarking Facial Image Analysis Technologies, Barcelona, pp. 2144–2151 (2011)
20. Burgos-Artizzu, X.P., Perona, P., Dollár, P.: Robust face landmark estimation under occlusion. In: ICCV 2013, Sydney, Australia (2013)
21. Sagonas, C., Tzimiropoulos, G., Zafeiriou, S., Pantic, M.: 300 faces in-the-wild challenge: the first facial landmark localization challenge. In: Proceedings of IEEE International Conference on Computer Vision (ICCV-W 2013), 300 Faces in-the-Wild Challenge (300-W), Sydney, Australia, pp. 397–403 (2013)
22. Hasan, K., Moalem, M., Pal, C.: Localizing facial keypoints with global descriptor search, neighbour alignment and locally linear models. In: IEEE International Conference on Computer Vision Workshops (ICCVW), Sydney, pp. 362–369 (2013)
23. Gross, R., Matthews, I., Cohn, J.F., Kanade, T., Baker, S.: Multi-PIE. In: Proceedings of The Eighth IEEE International Conference on Automatic Face and Gesture Recognition (2008)
24. Messer, K., Matas, J., Kittler, J., Jonsson, J.: XM2VTSDB: the extended M2VTS database. In: Proceedings of Audio and Video-based Biometric Person Authentication, pp. 72–77 (1999)

25. Císař, P., Železný, M., Krňoul, Z., Kanis, J., Zelinka, J., Müller, L.: Design and recording of Czech speech corpus for audio-visual continuous speech recognition. In: Proceedings of the Auditory-Visual Speech Processing International Conference 2005, Vancouver Island, pp. 1–4 (2005)
26. Giraudel, A., Carré, M., Mapelli, V., Kahn, J., Galibert, O., Quintard, L.: The REPERE corpus : a multimodal corpus for person recognition. In: Proceedings of the Eight International Conference on Language Resources and Evaluation (LREC 2012) (2012)

Attention Training Game with Aldebaran Robotics NAO and Brain-Computer Interface

Stepan Gomilko, Alina Zimina, and Evgeny Shandarov[✉]

Laboratory of Robotics and Artificial Intelligence,
Tomsk State University of Control Systems and Radioelectronics (TUSUR),
Tomsk, Russian Federation
evgenyshandarov@gmail.com

Abstract. This paper describes design, creation and preliminary testing of hardware and software for BCI (Brain-Computer interface)-based humanoid robot control. The system, the concept of which is presented in the article, is assumed to use for training ADHD patient's attention.

Keywords: Robot–human interaction · Aldebaran robotics NAO · Brain-computer interface

1 Introduction

Last years the very promising approaches in human-computer interaction (HCI) were investigated: speech recognition, movements, postures and gestures [1]. One of the new methods of HCI is brain-computer interface (BCI).

In recent years, BCI devices have become commercially available. Therefore, many new opportunities have become available to developers and researchers to create fundamentally new applications based on such devices.

One of such applications could be a hardware and software system that offers attention training programs for children with Attention Deficit Hyperactivity Disorder (ADHD). The prevalence of this psychological disorder makes it necessary to develop effective methods of therapy. One of the challenges researchers face is how to make a child genuinely interested in an object.

The main feature of this solution is its combination of the BCI and the humanoid robot NAO. The process of attention training is game-based.

2 Background

2.1 Attention Deficit Hyperactivity Disorder

ADHD encompasses a wide range of symptoms, but is most often associated with inattention and hyperactivity that interfere with one's ability to do everyday tasks. ADHD often leads to decreased performance in school during childhood, and difficulties working in adulthood.

© Springer International Publishing Switzerland 2016
A. Ronzhin et al. (Eds.): ICR 2016, LNAI 9812, pp. 27–31, 2016.
DOI: 10.1007/978-3-319-43955-6_4

According to the statistics, approximately 4 to 18 % of Russian children have ADHD, in the US this figure is 4–20 %, 1–3 % in the UK, 3–10 % in Italy, 1–13 % in China, 7–10 % in Australia. Boys are diagnosed with ADHD, approximately nine times more often than girls. It's important that ADHD symptoms are found not only in children, but also in adults. For example, 3–5 % of adults suffer from ADHD in the US. In 30–70 % of cases the ADHD symptoms remain for the whole life [2].

From the neurological point of view ADHD is a stable and chronic syndrome with no established treatment. However, most experts consider that the most effective approach is the combination of several methods that should be selected individually for each case. The methods of behavior modification, psychotherapy, educational and neuropsychological correction are used [3]. One of the methods of non-drug treatment for ADHD is a therapy based on the biofeedback (BFB). Biofeedback technique has been used to treat ADHD for over 30 years. Active participation of a child in treatment is one of the advantages of this method [4]. In the therapeutic phase a patient is taught the technique of brain activity self-correction using adaptive feedback where visualized EEG characteristics are the controlled parameter [5].

In 2012 Lim et al. showed that the 8-week training children through games using BCI led to a significant improvement in inattentive symptoms and hyperactive-impulsive symptoms. We decided to test whether this experiment efficiently if we add to the gameplay something bright, eye-catching. This element became a humanoid robot that looks like a child.

2.2 Global Software Engineering

From October 2013 and January 2014 Ritsumeikan University and Tomsk State University of Control Systems and Radioelectronics (TUSUR) held a collaborative Japanese-Russian university course in Global Software Engineering (GSE). Lectures were delivered by Victor Kryssanov (Ritsumeikan University), Tomasz Rutkowski (University of Tsukuba) and Evgeny Shandarov (TUSUR) via teleconference. The goal of the practical part of the course was to organize a group of students with different majors to execute the final project. Some of the students were located in Tomsk (Russia), and others were in Kyoto (Japan), so all discussions and collaboration within projects took place in social networks. During this course, students have been developing scenarios of human-robot interaction by BCI Emotiv EPOC. The robot used was Aldebaran Robotics NAO and it was located in Tomsk, meanwhile Emotiv EPOC was located in Kyoto.

As a result of its work, one of the groups has developed a scenario for treatment of ADHD in children. Our group: Titinunt Kitrungrotsakul, Dang Tuan Linh, Boonsita Roengsamut, Peeraphan Puttawetmongkol, Zhu Shuaizhen, Egor Sidorov, Gomilko Stepan, Dashkevich Mikhail.

2.3 Emotiv EPOC

Emotiv EPOC is a device that implements the BCI functions. It is a helmet with electrodes which read electroencephalography (EEG) of the user's mental activity. Emotiv EPOC has 14 sensors plus 2 references offer optimal positioning for accurate spatial resolution, gyroscope, Wi-Fi module and USB interface. Signals were analyzed by the BCI 2000. The P300 speller included in BCI 2000 is a BCI application with which it is possible to input a letter of the alphabet using human thought. P300ClassifierGUI is used for the analysis of collecting data together with P300 Speller [6].

This technology can be used for disabled patients, such as controlling an electric wheelchair, mind-keyboard, or playing a hands-free game. BCI based attention training game can be a potential new treatment for ADHD [7]. It has lower price than analogue and it has shorter preparation time to use [8]. However, before using the BCI individual calibration has to be carried out. It is important to ensure that the calibration is as quick as possible, because people with ADHD have wave P300 amplitude mostly lower than healthy persons, or extremely elevated. The amplitude of wave P300 correlates with lack of attention [9]. It should be noted that the signal pickup and command recognition requires 300 ms [6].

2.4 Aldebaran Robotics NAO

To hold the child's attention, we should involve him in gameplay and arouse his interest. For example, we can use a bright moving toy attracting attention. Motions should not repeat in the script.

The Aldebaran Robotics Nao was used for experiment. The Nao is a research platform used by more than 550 prestigious universities and research labs around the world. It is a small humanoid robot, measuring 58 cm in height, weighing 4.3 kg and having 25 degrees of freedom. It has a range of sensors and actuators: 2 loudspeakers, 4 microphones, 2 cameras, a gyroscope, an accelerometer, and range sensors (2 IR and 2 sonars). The robot has an embedded computational core and connects externally via Wi-Fi or Ethernet. The Nao has a generally friendly and non-threatening appearance, which is therefore particularly well suited for child-robot interaction [10].

3   Robot-Human Interaction Scenario and Experiment

Based on the review, the following robot-human interaction scenario was proposed. The BCI Emotive EPOC is used by child. The robot explains the rules of the game to the child by voice. The robot voice synthesized by ALTextToSpeech module of NAOqi framework. Then the robot demonstrates a sequence of four activities (e.g. go forward, backward, left, right) to the child. After that the robot asks the child to "repeat" the sequence of 'thinking' commands to the robot. Commands are formed by BCI. The robot executes these commands one by one and if the command is correct, the robot waits for the next command, and if not, it asks to try again. In the end, the robot congratulates a child and offers to play again.

The main goal of the proposed game is to train child to hold attention. Cycle of the game will completed successfully when the initial sequence of activities will fully performed by robot.

The diagram of the system is shown in Fig. 1. The BCI part of the work was carried out by students in Kyoto. The robot executing the commands was in Tomsk.

Communication between components of the system was conducted over HTTP on the public Internet networks. Client and server components communicate via a developed protocol. Protocol commands were transmitted to the server using the GET method in the format http://<server address>/set_command.php?<Command>. Command set included F (NAO moves forward), B (NAO moves back), R (NAO moves right), L (NAO moves left).

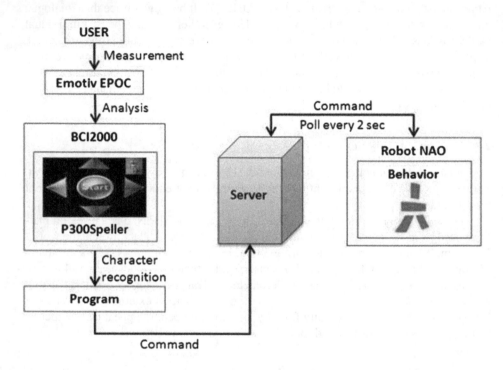

Fig. 1. Diagram of the system

The robot polls the server every two seconds for the command. The user interface is shown in Fig. 1 in P300 Speller section.

As we mentioned above, all work related to BCI carried out by our colleagues in Kyoto. Group in Tomsk has developed server software and robot software (Fig. 1). Choregraphe tool from Aldebaran Robotics was used to develope robot software and motions. This is a visual development environment that allows you to program robot and to create natural motions for NAO. We had used Python to develop some of parts of code. PHP was used to develop server software.

Preliminary testing of the system we carried out by hand forming commands. Collaborative experiment was in January 2014. Microsoft Skype was used to ensure visual contact research groups.

At the time of the experiment, the team has created five sequences of motions for the robot. Experiments were carried out for two hours. During this time, four complete cycles of the game have been successfully run. The main flaw of the system was a significant time interval between the beginning of the command formation and its execution by NAO, which was about 15–20 s. The main reasons for this were long delay in reading, analyzing and sending commands to the BCI (15–20 s) and delay in polling of the server by robot (2 s).

4   Conclusions

As a result of this work the team has proposed the system concept for training attention, implemented a prototype game and performed tests of the system via the public Internet networks. The main problem was the long delay between commands. Given the success of the experiments of Lim and others [7], this scenario may be used in further experiments with people suffering from ADHD. In the next work stage the therapeutic effectiveness of the prototype will be investigated.

References

1. Yusupov, R.M., Ronzhin, A.L.: From smart devices to smart space. Herald Russ. Acad. Sci. **80**(1), 45–51 (2010). MAIK Nauka
2. Zavadenko, N.N., Suvorina, N.U., Rumyantseva, M.W.: Hyperactivity, attention deficit: risk factors, age dynamics, diagnostic features. Defectology **6**, 13–20 (2003)
3. Perezhogin, L.O., Pokrovsky, V.I., Bochkov, N.P., Baranov, A.A., Belousov, U.B., Vaganov, N.N., Dmitrieva, T.B., Krasnov, V.N., Petruhin, A.S., Tiganov, A.C.: Attention Deficit Hyperactivity Disorder Cindrom (ADHD): Etiology, Pathogenesis, Clinical Course, Prognosis, Therapy, Assistance Organization. Expert Report, Moscow (2007)
4. Shtark, M.B.: EEG biofeedback with attention deficit hyperactivity disorder. Narcology **1**, 56–64 (2004)
5. Jafarova, O.A.: Game biocontrol as prevention technology of stress-dependent conditions. Biocontrol: Theor. Pract. **4**, 86–96 (2002)
6. Koike, T., Hayashi K., Ono, K., Ogawa, H.: Preliminary experiments with a brain-computer interface to control a robot. In: Innovations in Information and Communication Science and Technology, Tomsk (2013)
7. Lim, C.G., Lee, T.S., Guan, C., Fung, D.S.S., Zhao, Y., Teng, S.S.W., Zhang, H., Krishnan, K.R.R.: A brain-computer interface based attention training program for treating attention deficit hyperactivity disorder. PLoS ONE **7**(10), e46692 (2012)
8. Liu,Y., Jiang, X., Cao, T., Wan, F., Mak P.U., Mak, P.I., Vai, M.I.: Implementation of SSVEP Based BCI with Emotiv EPOC. IEEE (2012)
9. Efimiv, I.O., Ivanov, A.S., Nikitin, I.A.: Development and implementation of a comprehensive program of non-pharmacological rehabilitation of children with attention deficit hyperactivity disorder. Kazan Med. J. **92**(3), 379–382 (2011)
10. Nalin, M., Bergamini L., Giusti, A., Baroni, I., Sanna, A.: Children's perception of a robotic companion in a mildly constrained setting: how children within age 8–11 perceive a robotic companion. In: Proceeding of IEEE/ACM Human-Robot Interaction 2011 Conference (Robots with Children Workshop), Lausanne (2011)

Conceptual Model of Cyberphysical Environment Based on Collaborative Work of Distributed Means and Mobile Robots

Andrey Ronzhin[1]([⊠]), Anton Saveliev[1], Oleg Basov[1],
and Sergey Solyonyj[2]

[1] SPIIRAS, 39, 14th line, St. Petersburg 199178, Russia
ronzhin@iias.spb.su
[2] SUAI, 67, Bolshaya Morskaia str., St. Petersburg 190000, Russia

Abstract. In this paper, we propose a conceptual model of a cyberphysical environment based on a new approach to distribution of sensor, network, computing, information-control and service tasks between mobile robots, embedded devices, mobile client devices, stationary service equipment, and cloud computing and information resources. The task of structural-parametric synthesis of the corresponding cyberphysical system is formalized. Methods of integer-valued programming are used for the task solution.

Keywords: Cyberphysical environment · Cloud robotics · User behavior · Client devices · Data management · Proactive control

1 Introduction

Problems of two scientific paradigms, such as robotics and an ambient intelligent space, until recently were solved independently of each other. As a result, it was impossible to equip the robot with a multimodal user interface with the functions of a speech dialogue because many complex pattern recognition tasks could not be solved by limited computing and information resources of mobile robotic systems. Nowadays, with the development of cyberphysical systems and cloud robotics, as well as with the use of the so-called ecological approach, where a mobile robot only performs specialized functions that cannot be solved by stationary surrounding devices, the methods of human-robot communication and collaboration are changing dramatically.

Necessity to develop robotic companions capable of speech and multimodal communication with user is announced by developed countries, where the problem of aging and increasing number of incapable persons is very acute. Research activities of multimodal interfaces which offer natural communication between human and computer-based system were being held since end of 20th century. However, questions of single-modal analysis (speech, mimic or gesture, etc.) were being discussed since the middle of the 20th century. Development of the cloud technology and cyberphysical systems led to a breakthrough in the multimodal interfaces field. Application of distributed computing resources for collection, segmentation and analysis of large amounts of data from various modalities, which are used in conversations allows to

© Springer International Publishing Switzerland 2016
A. Ronzhin et al. (Eds.): ICR 2016, LNAI 9812, pp. 32–39, 2016.
DOI: 10.1007/978-3-319-43955-6_5

recognize audiovisual patterns. It can solve the problem of designing personalized multimodal interfaces, which sense the environment similarly to the user during the interaction.

An important point in the organization of functioning of such intelligent cybernetic spaces is the synthesis of the system for distributing tasks of ensuring multimodal interaction of users with mobile robots and with other service devices of cyberphysical systems (CPS).

2 Related Works

In [1], the authors raise the problem of the predictability of CPSs. In queuing systems that operate in real time, the accuracy and period of forecast become critical. In most cases, execution time of tasks and resources for solving these tasks are predicted using the measures of minimum/maximum tasks execution time [2].

The paper [3] is dedicated to the development of a crowd-sensor platform that employs user mobile devices to assess the social dynamics and delivery of personalized services. The new cloud service SAaaS (Sensing and Actuation as a Service) that performs for the user perception tasks and activities through mobile client devices and cloud resources is proposed in [4]. The problems solved by crowd-sensor mobile systems are presented in more detail in [5].

An overview of the basic principles, models and applications used in cyber-physical systems is presented in [6]. A future cyber-physical system, according to [7], must satisfy three basic principles of stability, security and consistency. Achieving these objectives is associated with the problems of concurrency and reliability of the operation of physical sensors, activators, and computer systems.

In [8], the authors discuss the possibility of using mobile robots to create a cascade communication network in remote areas or search and rescue regions where there is no cellular communications. With the use of the genetic algorithm and the method of partial swarm optimization, the calculation of coordinates of the position of individual robots and their peer-to-peer communications is conducted taking into account the location of obstacles.

An analysis of the existing approaches and technical solutions WiFI, ZigBee, RFID for passive determination of users in office premises with an extensive radiocommunications network infrastructure is presented in [9]. A distinctive feature of the proposed approach is the analysis of the change in the power of signals transmitted between wireless posts that occurs because of the presence of users on the signal propagation path. During this analysis no client devices are used. An evaluation of the system sensitivity to external factors and a comparison with other research systems were carried out in [10, 11]. In a similar study [12], the localization accuracy of up to 7 cm of five people moving in the room was experimentally obtained. An analysis of the existing wireless communication standards used in cyber-physical systems is presented in [13].

Despite the availability of the results of solving particular problems of cyber-physical systems management, currently there are no solutions to the problem of synthesis of such systems for specific applications and, as a consequence, to the important task of ensuring interaction between system elements. Conceptual modeling is an important step towards achieving these objectives.

3 Conceptual Model of Cyberphysical Intelligent Environment

Hereinafter, we will assume that a cyberphysical system has to solve a dual task:

First, it is creating in specified areas a physical space of such conditions ("information fields"), when each element (node) of CPS, located in these areas, can determine its location, share information with other elements, identify and assess the state of the environment (user).

Second, it is creating and maintaining such CPS structure that will ensure interaction (direct or energetic) with the user within a predetermined (or minimal) period of time; in the process of this interaction a target task facing the system will be executed.

It should be noted that the task of creating the above-mentioned conditions may be imposed both on the entire space (globally) and on any its part (locally); either on the entire predetermined time interval (continuously), or any discrete time instants and intervals (discretely). A set of numerical parameters that characterize certain conditions (information fields) may be additionally specified.

We consider mobile robotic, client, embedded, stationary and cloud components as the CPS elements (Fig. 1). Their main subsystems can be associated with four processes (types of functioning): interaction; functioning of target and service technical means; movement; consumption and (or) replenishment of resources.

It follows from the above that for each type of the CPS element we should formulate the aim of its functioning related to the processes of interaction with the user and other elements (nodes) and determine an appropriate sequence of actions to reach

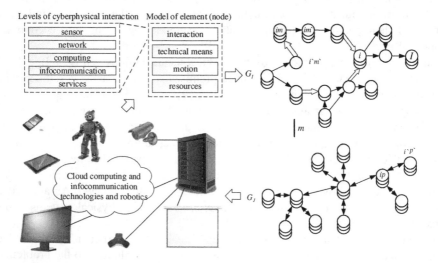

Fig. 1. A conceptual model of a cyberphysical space

the goal. An analysis shows that, in this case, it is convenient to conceptually describe the specified activity of CPS elements using the notion of "operation", by which we mean the action or system of actions having a common goal [14].

As is clear from Fig. 1, the purpose of functioning of the CPS element is realized when the element performs operations related to the information, material and energy exchange with the environment, the user, and other elements of the space. Thus, the exchange operation is a major, backbone factor, combining (integrating) various activities (movement, performance of technical means, resource consumption, etc.). Cyber-physical interaction is carried out at five main levels: sensor, network, computational, data management, service.

4 Task Distribution Between Elements of Cyberphysical Intelligent Environment

To formalize the interrelations between different variants of arranging the elements (or a plurality of such elements) of the system, alterative-graph formalization is used [15], in which the above-mentioned variants are given as peaks of an alternate graph, and the arcs reflect the nature of the interrelations between them (Fig. 1).

Let G_J be a graph defining the variants of the composition and the interrelation of possible CPS nodes; $G_J^* \in G_J$ is a subgraph that specifies one of the possible variants of the implementation of CPS nodes and interrelations between them. The vertices of the graph G_J are identified with variants of arranging the information processing nodes, possible locations of nodes, complexes of technical means, etc. The arcs of the graph reflect the interrelations between the nodes; G_I is a graph of the interrelations of performing alternative system management functions; $G_I^* \in G_I$ is a subgraph that defines one of the possible variants of implementation of system functions. The vertices of the graph G_I are identified with information processing procedures, management tasks and their stages, and so on depending on the problem. The arcs of the graph reflect levels of cyber-physical interaction. $\Re$ is a procedure of mapping the graph G_I upon G_J, which determines the distribution of functions, performed by the system, over its nodes. $n_\eta (\eta = \overline{1, \eta_0})$ is the quality characteristics of the creation and functioning of CPS.

Then the static problem of CPS structure synthesis can be represented as follows:

$$
\begin{aligned}
& \mathrm{extr}\Re_0 \{ (G_I^* \in G_I) \Re^* (G_J^* \in G_J) \}, \\
& \Re_\eta \{ (G_I^* \in G_I) \Re^* (G_J^* \in G_J) \}, \quad \eta = \overline{1, \eta_0}, \Re^* \in \Re.
\end{aligned}
\tag{1}
$$

In the allocation of tasks between the CPS nodes, two types of $\Re$ mappings are possible: (1) each task (stage) is executed only in one of the several possible system nodes; (2) tasks (stages) are executed in multiple system nodes.

Suppose we are given a plurality of tasks $(i = \overline{1, I})$ (stages $m = \overline{1, m_i}$) of interaction (sensor, network, computational, data management, service) and variants of their arrangement $(k = \overline{1, K})$, a plurality of system nodes $(j = \overline{1, J})$ and variants of their arrangement $(p = \overline{1, P_j})$.

The task of CPS structure synthesis, in this case, can be stated as follows:

$$F_0\left(x_{ik}, x_{imn}, x_{imj}, x_{jp}\right) \rightarrow \text{opt},\tag{2}$$

where F_0 are optimized quality indicators, e.g. a mean dwell time of tasks in the system; The expressions (3) and (4) determine the restriction on operating costs and on nodes loading respectively:

$$\sum_{i,m,n,j} B_{imnj\,i'm'n'j'} x_{imnj} x_{i'm'n'j'} \le B\tag{3}$$

$$\sum_{i,k,m,n} R^{\gamma}_{ikmnt} x_{ikmnj} \le R^{\gamma}_{jpt} - P^{\gamma}_{jpt}, \quad j = \overline{1,J}, \quad \gamma = \overline{1,\gamma_0},\tag{4}$$

where $B_{imnj\,i'm'n'j'} = \alpha_{imnj}$ if $imnj = i'm'n'j'$; $B_{imnj\,i'm'n'j'} = \beta_{imni'm'n'} \gamma_{jlj'l'}$, if $imnj \ne i'm'n'j'$; α_{imnj} are costs of performing the m-th stage of the i-th task in the j-th node; $\beta_{imni'm'n'}$ is the average flow of information between the mn-stage of the i-th task and m'n'-stage of the i'-th task in the process of system functioning; $\gamma_{ji'j'i'}$- costs of transmitting a unit of information from the node j to the node j` (the nodes are equipped with technical means of the l-th and l'-type, respectively); R^{γ}_{ikmnt} is the amount of resources of the γ-th type for time t needed to perform the m-th stage of the i-th task; P^{γ}_{jpt} are resources needed to carry out operational tasks.

$$x_{ik} = \begin{cases} 1, & \text{if the } i\text{-th task is solved using the } k\text{-th method;} \\ 0 & \text{- if not,} \end{cases}$$

is a variant of solving the task;

$$x_{imn} = \begin{cases} 1, & \textit{if the } m\text{-th stage of the } i\text{-th task is performed using the } n\text{-th method;} \\ 0 & \text{- if not,} \end{cases}$$

is a variant of performing the stage;

$$x_{imj} = \begin{cases} 1, & \textit{if the } m\text{-th stage of the } i\text{-th task is performed in the } j\text{-th node;} \\ 0 & \text{- if not,} \end{cases}$$

is a scope of the task;

$$x_{jp} = \begin{cases} 1, & \textit{if the } j\text{-th element is realized using the } p\text{-th method;} \\ 0 & \text{-if not,} \end{cases}$$

is a variant of node implementation;

$$x_{imnj} = \begin{cases} 1, & \text{if the } m\text{-th stage of the } i\text{-th task is performed in the } n\text{-th variant in the } j\text{-th node;} \\ 0 & \text{- if not} \end{cases}$$

is a variant of task solving, the stage and the node in which the stage is performed;

$$x_{ikmnj} = \begin{cases} 1, \text{ if the } n\text{-th variant of the } m\text{-th stage of the } i\text{-th task,} \\ \text{solved using the k-th method, is performed in the } j\text{-th node;} \\ 0 - \text{ if not,} \end{cases}$$

is a variant and a method of task solving, a variant of the stage and the node in which the stage is performed.

The presented formalization of the synthesis of the CPS structure (1) makes it possible to solve the problem of tasks distribution between the elements (nodes) of CPS using the methods of integer programming.

5   Conclusion

Solving the problem of synthesis of the optimal CPS structure is complicated taking into account the dynamics of user behavior, system elements and technical means. This is because the optimal rules for nodes functioning and the corresponding quality characteristics may be determined only after the list of problems solved by a given node as well as their optimal service procedure become known. In its turn, the distribution of tasks over the nodes depends on the characteristics of their maintenance in nodes.

Further research is expected to generate a list of tasks of multimodal service of users of cyberphysical intelligent environment [16–18]. For the implementation of the corresponding cyberphysical system, the minimum set of technical means is supposed to be used: 5 video cameras; 3 microphone arrays; 3 mobile client devices with different operating systems; 2 mobile robots; network equipment; activation executive devices; a server to support resource-intensive computations of audiovisual signal processing and the analysis of the external information resources required to obtain additional data on users. This technical means and technological platform of audiovisual signal processing were used for creation of an intelligent meeting room. Now they are being extended by using mobile robots and client devices to model collaborative work of distributed means and mobile robots for the user behavior analysis and provision of required services [19, 20].

Acknowledgment. The study was performed through the grant of the Russian Science Foundation (project № 16-19-00044).

References

1. Sun, B., Li, X., Wan, B., Wang, C., Zhou, X., Chen, X.: Definitions of predictability for cyber physical systems. J. Syst. Archit. (2016). doi:10.1016/j.sysarc.2016.01.007
2. Thiele, L., Wilhelm, R.: Design for timing predictability. Real-Time Syst. **28**, 157–177 (2004). Kirner, R., Puschner, P.: Time-predictable computing. In: Min, S.L., Pettit, R., Puschner, P., Ungerer, T. (eds.) SEUS 2010. LNCS, vol. 6399, pp. 23–34. Springer, Heidelberg (2011)

3. Merlino, G., Arkoulis, S., Distefano, S., Papagianni, C., Puliafito, A., Papavassiliou, S.: Mobile crowdsensing as a service: a platform for applications on top of sensing clouds. Future Gener. Comput. Syst. **56**, 623–639 (2016)
4. Distefano, S., Merlino, G., Puliafito, A.: Sensing and actuation as a service: a new development for clouds. In: Proceedings of the 2012 IEEE 11th International Symposium on Network Computing and Applications, NCA 2012, pp. 272–275. IEEE Computer Society, Washington, DC (2012)
5. Ganti, R., Ye, F., Lei, H.: Mobile crowdsensing: current state and future challenges. IEEE Commun. Mag. **49**(11), 32–39 (2011)
6. Hua, F., Lua, Y., Vasilakos, A., Haoc, Q., Maa, R., Patil, Y., Zhanga, T., Lua, J., Li, X., Xiong, N.: Robust cyber-physical systems: concept, models, and implementation. Future Gener. Comput. Syst. **56**, 449–475 (2016)
7. Hahn, A., Ashok, A., Sridhar, S., Govindarasu, M.: Cyber-physical security testbeds: architecture, application, and evaluation for smart grid. IEEE Trans. Smart Grid **4**(2), 847–855 (2013)
8. Mina, B., Kima, Y., Leea, S., Jin, J., Matsona, E.: Finding the optimal location and allocation of relay robots for building a rapid end-to-end wireless communication. Ad Hoc Netw. **39**, 23–44 (2016)
9. Gonga, L., Yanga, W., Zhoub, Z., Mana, D., Caic, H., Zhoud, X., Yange, Z.: An adaptive wireless passive human detection via fine-grained physical layer information. Ad Hoc Netw. **38**, 38–50 (2016)
10. Kosba, A.E., Saeed, A., Youssef, M.: RASID: a robust WLAN device-free passive motion detection system. In: Proceedings of IEEE International Conference on Pervasive Computing and Communications (PerCom), pp. 180–189 (2012)
11. Xiao, J., Wu, K., Yi, Y., Wang, L., Ni, L.: FIMD: fine-grained device-free motion detection. In: Proceedings of IEEE International Conference on Parallel and Distributed Systems (ICPADS), pp. 229–235 (2012)
12. Joshi, K., Bharadia, D., Kotaru, M., Katti, S.: WiDeo: fine-grained device-free motion tracing using RF backscatter. In: Proceedings of 12th USENIX Symposium on Networked Systems Design and Implementation (NSDI 2015), pp. 189–204 (2012)
13. Kabalci, Y.: A survey on smart metering and smart grid communication. Renew. Sustain. Energy Rev. **57**, 302–318 (2016)
14. Kalinin, V.N., Ohtilev, M., Sokolov, B.V.: Mul'tiagentnaja robototehnicheskaja inter-pretacija koncepcii aktivnogo podvizhnogo obyekta. Izvestija Kabardino-Balkarskogo nauchnogo centra RAN **6**(38), 148–157 (2010)
15. Cvirkun, A.D.: Osnovy sinteza struktury slozhnyh sistem. M.: Nauka (1982)
16. Ronzhin, A.L., Budkov, V.Y., Ronzhin, A.L.: User profile forming based on audiovisual situation analysis in smart meeting room. SPIIRAS Proceedings **23**, 482–494 (2012)
17. Basov, O.O., Struev, D.A., Ronzhin, A.L.: Synthesis of multi-service infocommunication systems with multimodal interfaces. In: Balandin, S., Andreev, S., Koucheryavy, Y. (eds.) NEW2AN/ruSMART 2015. LNCS, vol. 9247, pp. 128–139. Springer, Heidelberg (2015)
18. Ronzhin, A.L., Budkov, VYu.: Multimodal interaction with intelligent meeting room facilities from inside and outside. In: Balandin, S., Moltchanov, D., Koucheryavy, Y. (eds.) ruSMART 2009. LNCS, vol. 5764, pp. 77–88. Springer, Heidelberg (2009)

19. Karpov, A., Ronzhin, A., Kipyatkova, I.: An assistive bi-modal user interface integrating multi-channel speech recognition and computer vision. In: Jacko, J.A. (ed.) Human-Computer Interaction, Part II, HCII 2011. LNCS, vol. 6762, pp. 454–463. Springer, Heidelberg (2011)
20. Yusupov, R.M., Ronzhin, A.L.: From smart devices to smart space. Herald of the Russian Academy of Sciences, MAIK Nauka **80**(1), 45–51 (2010)

Control Method for Heterogeneous Vehicle Groups Control in Obstructed 2-D Environments

Viacheslav Pshikhopov, Mikhail Medvedev$^{(\boxtimes)}$, Anatoly Gaiduk, and Aleksandr Kolesnikov

Southern Federal University, Taganrog, Russia
pshichop@rambler.ru, medvmihal@sfedu.ru,
{gaiduk_2003,kolesnik7}@mail.ru

Abstract. The article considers the problem of distributed control for a group of heterogeneous vehicles. A survey of tasks and group control methods is given. A problem is posed to synthesize a local control algorithm ensuring motion if a heterogeneous group in a 2D environment with nonstationary obstacles. The algorithm is used to calculate the required speed and robot's heading. A principle is used that allows us to treat all the neighboring objects as repellers. Unlike the known methods, in the proposed approach the repelling forces are formed at the outputs of dynamic units allowing us to perform synthesis in the state space instead of a geometric space. Motion steady state modes analysis of the planned paths is performed and their stability is considered. The presented results allows to improve the operation of the robot safety among human environment.

Keywords: Heterogeneous groups · Group control · Vehicle · Decentralized control

1   Introduction

A group of robots usually joins robots supplementing each other's functionality [1]. So the potential of a group of robots is higher than that of an individual robot of any type. In group control problem, the groups often consist of intelligent robots, that are built using intelligent methods such as fuzzy logic, artificial neural networks and expert systems [2]. Intelligent control methods require performing mathematical and logical operations. So implementation of intellectual control methods requires high capabilities of the computing system.

In a number of cases task solution doesn't require action of the whole group of robots. In this case a subgroup of robots is organized that is called a "cluster" and is oriented on solution of a specific task [3].

Robot's group can implement the methods of centralized, decentralized or hybrid control strategy described in [4, 5] and other numerous articles. In centralized control methods the group of robots has a "robot-leader" Basing on information coming from the group member robots and on the information about the tasks set for the group by a

© Springer International Publishing Switzerland 2016
A. Ronzhin et al. (Eds.): ICR 2016, LNAI 9812, pp. 40–47, 2016.
DOI: 10.1007/978-3-319-43955-6_6

higher level control system, control system solves the tasks of cluster formation and distributes the tasks among them. A more promising approach is the decentralized control strategy that leads to distributed group control systems. In this case the group control system is implemented by informational junction of computing systems of several robots or all the robots in the group [4, 5].

Currently, the main issue is the safe operation of the robot in the human environment. In this context, this article deals with the problem of constructing a control system that ensures safe operation in an environment with obstacles.

For the first time the idea of using repelling and attracting sets in vehicles control was introduced in the works of Platonov in 1970 [6], where the potentials method was presented as a solution of the path finding problem. In the world literature the main references are made to the works of Brooks and Khatib [7, 8] published in 1985 and 1986. Another mobile robot control work using the force fields ideas was performed by Hitachi company in 1984 [9]. Nowadays the potential field method is widely spread. The work [10] presents the idea of transforming point obstacles into repellers using Lyapunov instability theorem. In [11] this approach is extended for 3D space and in [12] the control task is considered for environments where obstacles can take various configurations.

Nowadays potential field method is used widely. In [13] artificial potential field are applied for autonomous space mobile robots. Time-variant artificial potential fields are proposed. In [14] an enhanced potential field method that integrates Levenberg-Marquardt algorithm and k-trajectory algorithm into the basic potential field method is proposed and simulated.

2   Problem Statement

Let's consider vehicles with the following equations of kinematics (Fig. 1):

$$\dot{y}_{1i} = V_i \cos \varphi_i, \quad \dot{y}_{2i} = V_i \sin \varphi_i, \tag{1}$$

where y_{1i}, y_{2i}, V_i – vehicle's coordinates and speed; φ_i - heading angle; $i = \overline{1, n}$.

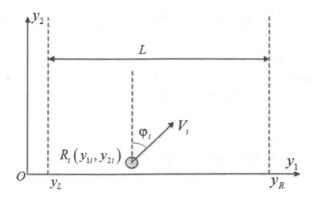

Fig. 1. State variables and coordinates system

Vehicle's position is described by coordinates y_{1i}, y_{2i} in the coordinates system Oy_1y_2. Speed V_i and heading φ_i are the controls.

3   Control Algorithm

Let $y_{2i} = 0$, and $y_{1i} \neq y_{1j}$, $\forall i \neq i$, $i,j = \overline{1,n}$. Let's enumerate the vehicles so that their index $i = \overline{1,n}$ increases with increasing coordinate y_{1i}.

Assume that the group consists of heterogeneous vehicles that require different distances among them. Let's introduce linear functions that will be used as a foundation for repellers. In Fig. 2, y_{1i} – position of ith vehicle; y_{1i-1} – position of a vehicle or obstacle next to the left; y_{1i+1} – position of a vehicle or obstacle next to the right.

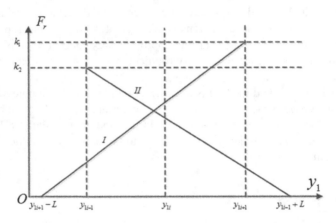

Fig. 2. Formation of repellers with linear repelling forces for a heterogeneous group of vehicles

Form the equations of line I, and II presented in Fig. 2 we get the equations for additional dynamic variables forming the repellers in the state space of the vehicles' group

$$\dot{z}_i = \frac{(k_{1i} + k_{2i})y_{1i} - k_{1i}y_{1i+1} - k_{2i}y_{1i-1} + (k_{1i} - k_{2i})L}{L}, \quad i = \overline{1,n}. \tag{2}$$

It is necessary to find controls φ_i, V_i ensuring stabilization of additional variables z_i and motion of robots along the axis Oy_2 with constant speeds. For solution of this task we introduce quadratic functions of the following form

$$V_i = 0.5z_i^2. \tag{3}$$

The derivative of the expression (3) accounting for the Eq. (2) is

$$\dot{V}_i = z_i \dot{z}_i = z_i((k_{1i} + k_{2i})y_{1i} - k_{1i}y_{1i+1} - k_{2i}y_{1i-1} + (k_{1i} - k_{2i})L)/L. \tag{4}$$

In order to ensure that the function (4) in negatively definite and guarantee constant motion speed the following functional relations are to be satisfied

$$\psi_i = \left[\begin{array}{c} \frac{(k_{1i}+k_{2i})y_{1i}-k_{1i}y_{1i+1}-k_{2i}y_{1i-1}+(k_{1i}-k_{2i})L}{L} + \alpha_i z_i = 0 \\ \dot{y}_{21} - V_k = 0, \ \ y_{2i} - y_{2i-1} = 0, \ i = \overline{2,n} \end{array} \right]. \tag{5}$$

Let's require the closed-loop loop system of the ith vehicle to satisfy the following reference equations

$$\dot{\psi}_i[1] + T_{1i}\psi_i[1] = 0, \ \ i = \overline{1,n}$$
$$\psi_1[2] = 0, \ \ \dot{\psi}_i[2] + T_{2i}\psi_i[2] = 0, \ \ i = \overline{2,n}. \tag{6}$$

From Eqs. (1), (5), (6) we get

$$\left[\begin{array}{c} u_{ix} \\ u_{iy} \end{array} \right] = \left[\begin{array}{c} \frac{L}{k_{1i}+k_{2i}} \left(\frac{k_{1i}\dot{y}_{1i+1}+k_{2i}\dot{y}_{1i-1}}{L} - \alpha_i \frac{(k_{1i}+k_{2i})y_{1i}-k_{1i}y_{1i+1}-k_{2i}y_{1i-1}+(k_{1i}-k_{2i})L}{L} \right) \\ -\frac{LT_{1i}}{k_{1i}+k_{2i}} \left(\frac{(k_{1i}+k_{2i})y_{1i}-k_{1i}y_{1i+1}-k_{2i}y_{1i-1}+(k_{1i}-k_{2i})L}{L} + \alpha_i z_i \right) \\ \left\{ \begin{array}{l} V_k, \ i = 1 \\ \dot{y}_{2i-1} - T_{2i}(y_{2i} - y_{2i-1}), \ i = \overline{2,n} \end{array} \right. \end{array} \right], \tag{7}$$

$$V_i = \sqrt{u_{ix}^2 + u_{iy}^2}, \ \ \varphi_i = \arctan\left(u_{iy}/u_{ix}\right). \tag{8}$$

Accounting for (8) the equations of the closed-loop system have the following form

$$\left[\begin{array}{c} \dot{y}_{1i} \\ \dot{y}_{2i} \\ \dot{z}_i \end{array} \right] = \left[\begin{array}{c} u_{ix} \\ u_{iy} \\ \frac{(k_{1i}+k_{2i})y_{1i}-k_{1i}y_{1i+1}-k_{2i}y_{1i-1}+(k_{1i}-k_{2i})L}{L} \end{array} \right]. \tag{9}$$

The closed-loop system (9) decomposes into two independent subsystems consisting of the first and third; and the second equations. So let's perform separate analysis of these subsystems. At first let's consider a subsystem consisting of the first and the third equations of the system (9).

Expressions for the steady state for system (11) are:

$$y_{1i} = \frac{k_{1i}y_{1i+1}+k_{2i}y_{1i-1}-(k_{1i}-k_{2i})L}{k_{1i}+k_{2i}}, \ \ z_i = 0, \ \ i = \overline{1,n}. \tag{10}$$

For i^{th} robot the variables y_{1i+1}, y_{1i-1} are external, i.e. the control systems of each robot are autonomous. Besides, since system (10) is linear, zero equilibrium point is analyzed. Thus stability analysis comes down to analysis of the following system:

$$\begin{bmatrix} \dot{y}_{1i} \\ \dot{y}_{2i} \\ \dot{z}_i \end{bmatrix} = \begin{bmatrix} -(\alpha_i + T_{1i})y_{1i} - LT_{1i}\alpha_i z_i/(k_{1i} + k_{2i}) \\ \begin{cases} V_k, \ i = 1 \\ -T_{2i}y_{2i}, \ i = \overline{2, n} \end{cases} \\ (k_{1i} + k_{2i})y_{1i}/L \end{bmatrix}. \tag{11}$$

The stability conditions of the system (11) have the following form

$$\alpha_i > 0, \ T_{1i} > 0, \ T_{2i} > 0. \tag{12}$$

Figure 3 presents the modeling results for the closed-loop system (11). The parameters are $\alpha_i = 2$, $T_{1i} = 3$, $T_{2i} = 3$. The number of robots is equal to 5. Without any obstacles in the robot's working area $k_{ji} = L$, $j = 1, 2$; $i = \overline{1, 5}$. If an obstacle emerges, the robots closest to it change the repulsion coefficient. In our example $k_{12} = 1.3L$, $k_{23} = 1.3L$.

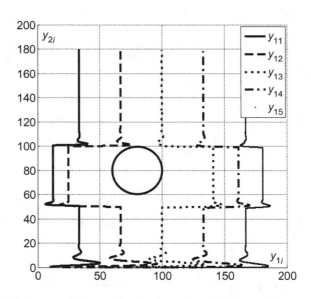

Fig. 3. System modeling results for piecewise-constant repulsion coefficients

4 Analysis of Control Algorithm in an Environment with Mobile Obstacles

For the sake of simplicity, analysis is performed for the case $k_{ji} = k$. However, all the reasoning can be applied to the case of a homogenous group of vehicles. Assume that obstacles motion speeds are constant. At a certain moment of time t_1 the obstacle gets into the robot's working area. At the moment $t_2 > t_1$ the obstacle gets out of this area. Let's assign $k = L$, $\alpha_i = k/L$. The variables y_{1i+1}, y_{1i-1} are considered to be external for an ith robot and obstacles' motion speeds are constant. From (11) we get the following system

$$\begin{bmatrix} \dot{z}_i \\ \dot{y}_{1i} \end{bmatrix} = \begin{bmatrix} 0 & 2k/L \\ -0.5T_{1i} & -k/L - T_{1i} \end{bmatrix} \begin{bmatrix} z_i \\ y_{1i} \end{bmatrix} + \begin{bmatrix} A_1 t + A_2 \\ A_3 t + A_4 \end{bmatrix}. \tag{13}$$

$A_1 = 0.5(k/L + T_{1i})(V_{i-1} + V_{i+1})$, $A_2 = 0.5(V_{i-1} + V_{i+1}) + 0.5(k/L + T_{1i})(V_{i-1}^0 + V_{i+1}^0)$,
$A_3 = -k(V_{i-1} + V_{i+1})/L$, $A_4 = -k(V_{i-1}^0 + V_{i+1}^0)/L$

$$\dot{y}_{1i-1} = V_{i-1}, \ y_{1i-1} = V_{i-1}t + V_{i-1}^0, \quad \dot{y}_{1i+1} = V_{i+1}, \ y_{1i+1} = V_{i+1}t + V_{i+1}^0,$$

where V_{i-1}, V_{i-1}^0, V_{i+1}, V_{i+1}^0 - measured constants.
   Solving system (13) we get:

$$\begin{cases} z_i = C_1 e^{-\frac{k}{L}t} + C_2 e^{-T_{1i}t}; \\ y_{1i} = -0.5C_1 e^{-\frac{k}{L}t} - 0.5C_2 T_{1i} L e^{-T_{1i}t}/k + 0.5(V_{i-1} + V_{i+1})t + 0.5(V_{i-1}^0 + V_{i+1}^0). \end{cases} \tag{14}$$

The free component of the expression (14) approaches zero if the values of k/L, T_{1i} are positive. The forced component leads to the change of the robot's position by the following value

$$\Delta y_{1i} = 0.5(V_{i-1} + V_{i+1})(t_2 - t_1). \tag{15}$$

Let's consider a situation presented in Fig. 4. Suppose that in steady state a group of robots moves with a constant speed V_k directed along the axis Oy_2. The transverse speed components of the robots become equal to zero at the time the robot reaches the fracture line of the left boundary. In the proposed approach a shift of the working area boundary is interpreted by the control system as a mobile obstacle moving with a constant speed. Let robots moving with a speed V_k pass through the segment BC during the time $(t_2 - t_1)$. Then from the triangle ABC we get:

$$V_L = V_k ctg \gamma. \tag{16}$$

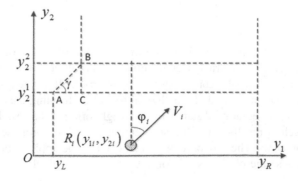

Fig. 4. Motion of a robot in the area with variable boundaries

Now let's consider situation when a mobile obstacle emerges between i^{th} and $(i-1)^{th}$ robots during the time interval $(t_2 - t_1)$. In this case the group of robots splits into 2 subgroups having obstacles at both sides. Let's consider one the subgroups consisting, e.g., out of robots numbered from 1st to i-1. In this case we can get the expression:

$$\Delta \bar{y}_{1j} = \left[(i-j+1)V_L + jV_{1p}\right]\Delta t/(i+1), \quad j = 1, i-1. \tag{17}$$

Figure 5 presents the modeling results for a group consisting of five robots. The modeling conditions match the ones of the previous example. The left boundary of the working area is described by the following equation:

$$y_L = \begin{cases} 0, & \forall(y_2 > 100) \ \&(y_2 < 50) \\ 0.5t, & \forall(50 \le y_2 \le 100) \end{cases}, \quad j = 1, i-1,$$

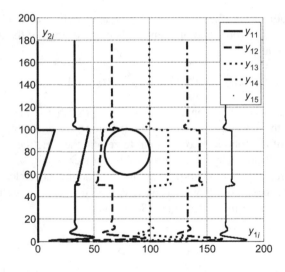

Fig. 5. Motion of the robots group in the area with variable boundaries

5 Conclusion

The article proposes and analyzes an algorithm of distributed control of a group of heterogeneous vehicles in obstructed environment. The algorithm is built using the control principle that allows the control system to interpret all the neighboring objects as repellers. We propose a method of repellers introduction forming the repelling forces using a dynamic unit integrating distance to the neighboring obstacles. The performed analysis and modeling results demonstrate effectiveness of the proposed methods for obstructed environments. The proposed approach can also be applied for nonstationary environments because the obstacles are formally treated as vehicles.

The proposed algorithms can be used in path planning systems of different type of vehicles [15, 16].

Acknowledgements. This work was supported by a grant from the Russian Science Foundation 14-19-01533 performed Southern Federal University.

References

1. Parker, L.E., Tang, F.: Building multirobot coalitions through automated task solution synthesis. Proc. IEEE **94**(7), 1289–1304 (2006)
2. Veres, S.M., Molnar, L., Lincoln, N.K., Morice, C.P.: Autonomous vehicle control systems — a review of decision making. Proc. Inst. Mech. Eng. Part I: J. Syst. Control Eng. **225**(3), 155–195 (2011)
3. Kalyaev, I.A., Gaiduk, A.R., Kapustyan, S.G.: Models and Algorithms of Collective Control of Robot Groups. Fizmatlit, Moscow (2009). (In Russian)
4. Tanner, H.G., Christodoulakis, D.K.: Decentralized cooperative control of heterogeneous vehicle groups. Robot. Auton. Syst. **55**(11), 811–823 (2007)
5. Ivchenko, V.D., Korneev, A.A.: Analysis of tasks distribution in the problem of robots group control. Mechatron. Autom. control **7**, 36–42 (2009)
6. Platonov, A.K., Kirilchenko, A.A., Kolganov, M.A.: Potential Method in Problem of the Path Planning: background and Perspectives. Institute of Applied Mathematics of M.V. Keldysh RAS, Moscow (2001)
7. Khatib, O.: Real-time obstacles avoidance for manipulators and mobile robots. Int. J. Robot. Res. **5**(1), 90–98 (1986)
8. Brooks, R.A.: Self calibration of motion and stereo vision for mo-bile robots. IEEE Int., Robotics and Automation (1986)
9. Ichikawa, Y., Fujie, M., Ozaki, N.: On mobility and autonomous properties of mobile robots. Robot **44**, 31–36 (1984)
10. Pshikhopov, VKh: Repellers forming in the process of mobile robots movements in environment with obstacles. Mechatron. Autom. Control **2**, 34–41 (2008)
11. Pshikhopov, VKh, Medvedev, MYu., Krukhmalev, V.A.: Position-path control of vehicles in the three dimensional environment with point obstacles. Izv. SFEDU. Tech. Sci. **1**(162), 238–250 (2015)
12. Beloglazov D.A., Guzik V.F., Kosenko E.Y., Krukhmalev V.A., Medvedev M.Y., et. al.: Intelligent planning of vehicles path in the environment with obstacles. Edited by V.K. Pshikhopov, Moscow, FIZMATLIT (2014)
13. Macktoobian, M., Shoorehdeli, M.A.: Time-variant artificial potential field (TAPF): a breakthrough in power-optimized motion planning of autonomous space mobile robots. Robotica **34**(5), 1128–1150 (2016)
14. Li, G., Chou, W.: An improved potential field method for mobile robot navigation. High Technol. Lett. **22**, 16–23 (2016)
15. Pshikhopov, V.K., Medvedev, M.Y., Gaiduk, A.R., Gurenko, B.V.: Control system design for autonomous underwater vehicle. In: Proceedings IEEE Latin American Robotics Symposium, LARS 2013, pp. 77–82 (2013)
16. Pshikhopov, V.K., Medvedev, M.Y., Gaiduk, A.R., Fedorenko, R.V., Krukhmalev, V.A., Gurenko, B.V.: Position-trajectory control system for unmanned robotic airship. In: IFAC Proceedings Volumes (IFAC-PapersOnline), pp. 8953–8958 (2014)

Decentralized Approach to Control of Robot Groups During Execution of the Task Flow

Igor Kalyaev, Anatoly Kalyaev[✉], and Iakov Korovin

Scientific Research Institute of Multiprocessor Computer System,
Southern Federal University, Taganrog, Russia
anatoly@klayaev.net, korovin_yakov@mail.ru

Abstract. The current paper covers the solution of the problem of decentralized control of a network-centric group of heterogeneous robots (robots of various types), that have various functional capabilities, during collective execution of complicated tasks received from consumers in a priori unknown moments of time. Here, the complicated task is the task which requires performance of a certain set of various interconnected operations for its execution. In the paper we describe methods of multi-agent adaptive distribution of operation of complicated tasks between the robots of the network-centric group according to their current condition and functional specialization. The methods are based on interaction of program agents which represent interests of the individual robots of the group during distribution of operations of the incoming tasks.

Keywords: Network-centric group of robots · Heterogeneous robots · Complicated task · Decentralized control · Multi-agent distribution of operation · Program agent · Prototype of program service · Model of robotics group

1 Introduction

A robot is a technical system which is capable to substitute human during execution of certain operations, and first of all those, that can cause damage or death to people during execution. During last decades robots are used in many domains of science and techniques, particularly in space and underwater research, or take part in removing aftermath of emergency situations and anthropogenic disasters, guarding of territories and research of area, military operations and etc. However, at present it becomes more and more evident that a single robot is capable to solve a rather limited class of tasks. At the same time solution of complicated tasks is possible only if we use a group [1] of robots with various functional capabilities. The principal advantages of group use of robots during of execution of a complicated task are:

1. Reduction of the task execution time because it is possible to distribute its separate operations between the robots.
2. Increasing of probability of successful execution of the task because failure of one individual robot does not lead to failure of the whole system, and operations assigned to this robot can be re-distributed between the rest of the robots of the system.

© Springer International Publishing Switzerland 2016
A. Ronzhin et al. (Eds.): ICR 2016, LNAI 9812, pp. 48–54, 2016.
DOI: 10.1007/978-3-319-43955-6_7

2   Current State of Research

At present research in the domain of group use of robots are actively undertaken in many advanced countries of the world [2–4]. According to analysis of results of research in the domain of group behavior of robots we can make the following conclusions:

1. There is no general methodology of task solution, when robots are controlled as a group. Each time the developer team has practically to invent and develop new algorithm of group control that can be used for solution of their particular tasks.
2. In the majority of projects developers use principles of either centralized control of the robot group, when control is provided by a single centre, or aggregative control, when behavior of the robots of the group is defined by elementary rules of interaction with neighbor robots. The first approach does not allow control of large groups of robots because one centralized control node (CCN) cannot provide effective service in a real-time mode and has low fault tolerance, because failure of the CCN or communication lines has disastrous consequences for the whole group, and the second approach does not provide effective execution of complicated tasks which require interaction of the robots of the group for concurrent execution of various actions.

That is why the problem of development of a certain generalized methodology of group control of robotics groups during execution of complicated tasks is extremely urgent. The methodology is based on use of unified software tools which allow creation of a network-centric group of robots and unlimited scaling of its members, and decentralized control of the group during execution of the flow of a priori unknown user tasks.

3   The Decentralized Approach

Owing to decentralized network-centric organization of the robotics group, it is possible to avoid the specified shortcomings. Here we assume that each robot of the group has its own individual control device (CD) and coordination of their collective actions during execution of the group task is performed by means of information interaction via some communication bus [5, 6]. Such network-centric organization of the robot group provides: first of all, high fault-tolerance of system because it has no "bottleneck" like CCN, and failure of any robot does not lead to failure of the whole group; secondly, ability of practically unlimited increasing on the number of the robots in the group by simple connection to the communication bus and, finally, reduction of computational load of the control device of an individual robot because it must control the given robot but not the whole group. In turn, it reduces technical requirements to the CCN and it provides real-time control.

However, on the other hand, such decentralized organization of a network-centric robotics group requires development of fundamentally new methods and algorithms of group control of robots during execution of complicated tasks [7].

4 Problem Definition

In a generalized form the problem of control of a network-centric group consisted of robots of various types can be presented as follows.

Let us assume that a network-centric group $\mathbf{R}$ contains N robots $R_1, R_2, \ldots R_N$, where each robot can perform a certain set of operations $\mathbf{A}_i =< A_1^i, A_2^i, \ldots A_L^i >$ $(i = 1, 2, \ldots, N)$. Let us consider that the robots of the group must perform a certain set (a flow) of various complicated tasks $\mathbf{Z} =< Z_1, Z_2, \ldots, Z_M >$, which can be received from different consumers in random moments of time. Here each such complicated task $Z_l \in \mathbf{Z}$ is represented as an acyclic graph of operations $\mathbf{G}_l(\mathbf{Q}_l, \mathbf{X}_l)$, whose vertex $q_j \in \mathbf{Q}_l$ corresponds to an operation A_j, which belongs to the set $\mathbf{A} = \bigcup_{i=1}^{N} \mathbf{A}_i$, and the edge $x(q_j, q_{j+1})$ means, that the result of the operation A_j, which corresponds to the vertex q_j, is required for the operation A_{j+1}, which corresponds to the vertex q_{j+1}. Besides, the consumer defines the time moment $T_{\max}^l$, in which he would like to receive the result of his task Z_l.

The aim of the robots of the network-centric group $\mathbf{R}$ is to perform all tasks of consumers to the specified moments of time [8].

5 Suggested Approaches

It is obvious that the problem of execution of the user's task Z_l by the group of robots $\mathbf{R}$ can be solved in two steps:

1. First of all, among the whole set of robots it is necessary to select a subset (an association) of robots, which provide execution of the specified task to the required moment of time $T_{\max}^l$, to distribute the operations that correspond to the vertices of the graph $\mathbf{G}_l(\mathbf{Q}_l, \mathbf{X}_l)$ of the task Z_l between the robots of the association and to set a time schedule of execution of the interconnected operations.
2. After that each robot $R_i \in \mathbf{R}$ of the association, performs all its operations according to the time schedule.

It is evident, that here the main problem is the first step [9]. To be more specific, it is the problem of distribution of the operations of the task between the robots of the group and specification of the time schedule of their execution. The problem of specification of the time schedule (network planning) was analysed in many research works. Therefore, if the graphs $\mathbf{G}_l(\mathbf{Q}_l, \mathbf{X}_l)$ $(l = 1, 2, \ldots, M)$ of all tasks of the set $\mathbf{Z}$ are defined and the group of robots $\mathbf{R}$ remains unchanged, then, using well-known methods, we can beforehand specify such schedules for each task from the set $Z_l \in \mathbf{Z}$ and save them in the memory of robots R_i ($i = \overline{1, N}$). Then, when a new task Z_l is received, the CD of each robot of the group can read from the memory the distribution (schedule) of operations which correspond to the task and start to execute them.

However, this is the ideal case. In reality graphs of operations of performed tasks can be a priori unknown, and the cast of the group can be unpredictably changed. For example, failures of some robots are possible [10]. Besides, the tasks, as we assumed before, can income in a priori unknown moments of time, and therefore it is impossible to use ready-made schedules for their execution, because the robots, which are required for execution of the new task, can be involved, in the current moment of time, in execution of the tasks which were received earlier. Therefore, we have a problem of development of the method of adaptive distribution (making of schedules) of operations for all robots of the group $\mathbf{R}$ during execution of the flow of a priori unknown tasks $\mathbf{Z}$ with the help of the control devices of individual robots of the group.

The problem of schedule making can have four different statements, that depend on the specified organization of the network-centric group:

1. In the most simple case all robots of the group $\mathbf{R}$ are similar and are able to perform similar sets of operation, i.e. $\mathbf{A}_i = \mathbf{A}_j, (i = 1, 2, \ldots, N, j = 1, 2, \ldots, i - 1, i + 1, \ldots, N)$. In addition, similar operations $A_s \in \mathbf{A}_i$ are performed by all robots during the same time, i.e. $t_i(A_s) = t_j(A_s)$ $(i = 1, 2, \ldots, N, j = 1, 2, \ldots, i - 1, i + 1, \ldots, N)$. Such case corresponds to a completely homogeneous group of robots that have the similar specialization and similar functional capabilities for execution of similar operations.

2. In the more complex case sets of operations performed by individual robots of the group $\mathbf{R}$, are similar, i.e. $\mathbf{A}_i = \mathbf{A}_j (i = 1, 2, \ldots, N, j = 1, 2, \ldots, i - 1, i + 1, \ldots, N)$, but the time of execution of identical operations $A_s \in \mathbf{A}_i$ by different robots of the group $\mathbf{R}$ is different. It depends on the status of the robot and on the conditions of the environment, i.e. $t_i(A_s) \neq t_j(A_s)(i = 1, 2, \ldots, N, j = 1, 2, \ldots, i - 1, i + 1, \ldots, N)$. This case corresponds to a homogeneous group of robots with similar specialization and different functional capabilities of execution of similar operations.

3. In the third case the sets of operations, performed by different robots of the group $\mathbf{R}$, are also different, i.e. $\mathbf{A}_i \neq \mathbf{A}_j$ $(i = 1, 2, \ldots, N, j = 1, 2, \ldots, i - 1, i + 1, \ldots, N)$, though it is also possible that $\mathbf{A}_i \cap \mathbf{A}_j \neq O$. Here the time of execution of similar operations $A_s \in \mathbf{A}_i$ for all robots $R_i (i = \overline{1, N})$ is the same, i.e. $t_i(A_s) = t_j(A_s)$ $(i = 1, 2, \ldots, N, j = 1, 2, \ldots, i - 1, i + 1, \ldots, N)$. The case corresponds to a heterogeneous group of robots, when each robot has its own specialization, but their technical capabilities of execution of similar operations are similar.

4. Finally, in the most complex case the robots of the group $\mathbf{R}$ perform different sets of operations, i.e. $\mathbf{A}_i \neq \mathbf{A}_j$ and $\mathbf{A}_i \cap \mathbf{A}_j \neq O$, and the time of execution of similar operations $A_s \in \mathbf{A}_i$ by different robots is also different, i.e. $t_i(A_s) = t_j(A_s)$ $(i = 1, 2, \ldots, N, j = 1, 2, \ldots, i - 1, i + 1, \ldots, N)$. This case corresponds to completely heterogeneous group of robots in which all robots have different specialization and have different functional capabilities for execution of similar operations.

6   Methods of Multi-agent Adaptive Distribution of Operations of Complicated Tasks Between the Robots of the Network-Centric Group

The first question, which arises when we are trying to the problem of control of network-centric group of robots, is how and in what form the robots must receive the tasks from consumers. It can be provided by a certain specially dedicated node connected to the common communication channel, i.e. so-called "bulletin board" (BB) which can be used for placement of consumer tasks. So, before placing the task Z_l on the BB, the consumer is to represent it in some formalized form – as a descriptor. The descriptor of the task Z_l must contain the following data:

- the graph $G_l(Q_l, X_l)$ of the task Z_l, represented, for example, as a table of complexity;
- the list of vertices of the set Q_l and operations of the set A, that correspond to the vertices;
- the time moment T_{max}^l, to which the task must be executed.

The second question is how the robots must distribute the operations that correspond to the vertices of the incoming graphs of the tasks, and specify the schedule of their execution. It can be provided by means of the multi-agent approach [7, 8]. Here we suggest to use program agents as active elements. They are physically implemented within the CD of the each robot $R_i \in R(i = 1, N)$ of the group which "represents its interests" during distribution and execution of the consumer's tasks. It is possible to represent functioning of the network-centric group of robots during execution of the flow of consumers' tasks as follows.

1. The user forms his task $Z_l \in Z$ as a graph $G_l(Q_l, X_l)$ and defines the required moment of time T_{max}^l, to which the solution must be received. The descriptor of the task $Z_l \in Z$ represented in such a form is placed on one of bulletin boards connected to the network-centric infrastructure.

2. The program agent of the robot R_i, which is not involved into execution of any other tasks, polls bulletin boards, searching the job for "its" robot. If a certain task $Z_l \in Z$ is found on the BB, then the agent tries to become a member of the association R_l which will execute the task. So, the agent of the robot R_i selects a fragment $Z_l \in Z$ in the graph of the task, that is the most complicated and takes the longest time to be executed, i.e. a certain sequence of vertices not occupied by other robots, and that can be executed to the specified moment of time by a corresponded robot.

3. If such fragment is found, then the agent R_i joins the association R_l that executes this task. The agent removes vertices that correspond to the selected fragments from the graph of the task and assigns the required execution time to the adjacent tops of the remaining subgraph. The execution time is calculated in order to provide execution of the operations of the selected fragment to the specified moment of time.

4. Then the robot R_i starts execution of the operations which correspond to the task fragment selected by the agent. Performing each operation of the fragment the agent of the robot checks, first of all, if all necessary data are ready, and controls the time schedule, estimating the time of execution of the whole fragment. If the initial data are still not received from the other members of the association, then the robot goes into the waiting mode. If the robot cannot provide execution of the task fragment in the specified time, i.e. it cannot follow the time schedule, then the agent makes a report on the bulletin board and leaves the association.
5. When the robot R_i has successfully executed the selected fragment of the task Z_l, its agent starts polling the bulletin boards again and tries to find a new task for its robot.

7 The Experiments and the Results

Verifying the efficiency of proposed algorithms is a very hard task because there is great amount of parameters in such system and it is almost impossible to carry out experiments with real hardware. That is why we had to create the program model, which allows modeling group of up to 1000 robots solving up to 250 tasks [8].

The main target of experimental research is the practical proof of efficiency of proposed algorithms, but also we wanted to evaluate of the relative loss of time due to decentralized dispatching. That is why we had to carry out several series of experiments with different parameters.

There are many important parameters in network-centric group that is why we decided to use planning of experiments: we divided parameters to three groups: Primary (system parameters) (P), Secondary (parameters of specific tasks and robots) (S), and Tertiary (user-conditional parameters) (T). In the end, we decided to explore influence of following parameters of distributed CS:

- number of robots (P1);
- frequency of incoming tasks (P2);
- performance of robots (S1);
- complexity of task (S2);
- priority (T1);
- number of subtasks (T2);
- time for solving the task (T3).

For primary parameters we decided to use absolute values, for secondary parameters relative values and for tertiary parameters limited random values. To make result more precise due to random values we carried out every experiment three times.

The results show that the percent of actual load at robots was good: from 52 % to 90 %. That means that time loss due to organization ranged from 10 % to 48 % of the ideal time.

8 Conclusion

Implementation of the suggested approach provides:

– quasioptimal distribution of operations of various tasks between the robots of the group;
– high useful loading of the robots of the group during execution of the flow of user tasks;
– high probability of execution of the user tasks to the specified time;
– unlimited increasing (scalability) of the robot number of the group;
– increased fault-tolerance of the robot group because failure of any robot does not lead to failure of the whole system.

Acknowledgments. The paper is published due to financial support of the Russian Ministry of Education and Science, the project № 16 within the base part of the state task 2014/174 and due the financial support of RFBR, projects 15-37-20821, 16-29-04194 ofi-m.

References

1. Kaliaev, A.: Multiagent approach for building distributed adaptive computing system. Proc. Comput. Sci. **18**, 2193–2202 (2013)
2. Cao, Y., Fukunaga, A., Kahng, A.: Cooperative mobile robotics: antecedents and directions. Auton. Robot. **4**, 7–23 (1997)
3. Kingston, D., Beard, R.W., Holt, R.S.: Decentralized perimeter surveillance using a team of UAVs. IEEE Trans. Robot. **24**, 1394–1404 (2008)
4. Dorigo, M., Birattari, M.: Swarm intelligence. Scholarpedia, **2**(9) (2007)
5. Korovin, Y.S., Kalyaev, A.I.: Methods and algorithms of improving the efficiency of data transmission systems in oil corporations' enterprise networks. Oil Industry, pp. 96–100 (2013). (in Russian)
6. Korovin, I.S., Khisamutdinov, M.V., Kaliaev, A.I.: The application of evolutionary algorithms in the artificial neural network training process for the oilfield equipment malfunctions' forecasting. In: Proceedings of the 2nd International Symposium on Computer, Communication, Control and Automation, Advances in Intelligent Systems Research, vol. 68, pp. 253–257 (2013)
7. Kalyaev, A., Korovin, I.: New method to use idle personal computers for solving coherent tasks. In: AASRI Conference on Circuit and Signal Processing (CSP 2014), AASRI Procedia, vol. 9, pp. 131–137 (2014)
8. Kalyaev, A., Korovin, I.: Adaptive multiagent organization of the distributed computations. In: 2nd AASRI Conference on Computational Intelligence and Bioinformatics, AASRI Procedia, vol. 6, pp. 49–58 (2014)
9. Korovin, I.S., Kalyaev, A.I., Khisamutdinov, M.V.: The decentralized approach to network-centric management of oilfield. Oil Industry (2015)
10. Kalyaev, A., Korovin, I., Khisamutdinov, M., Schaefer, G., Ahad, M.A.R.: A novel method of organisation of a software defined network control system. In: 4th International Conference on Informatics, Electronics and Vision, ICIEV (2015)

Development of Wireless Charging Robot for Indoor Environment Based on Probabilistic Roadmap

Yi-Shiun Wu, Chi-Wei Chen, and Hooman Samani[✉]

Department of Electrical Engineering, National Taipei University, Taipei, Taiwan
{s410287005,s410287017,hooman}@mail.ntpu.edu.tw

Abstract. The aim of this paper is to develop a robotic system which can navigate in an indoor environment and charge several electrical devices. Several algorithms such as Travel Salesman Problem, Probabilistic Roadmap and Fuzzy C-means Clustering are used for development of such a system. The testbed is constructed by Arduino Uno, Arduino WIFI Shield, Go-between Shield by Mayhew Lab and Polulu Zumo robot for Arduino Ver.1.2. All the algorithms are coded by MathWorks MATLAB and Simulink. The core of the wireless charging robot is to optimize the best performance for single robot to charge multiple devices. Owing to the computation restriction of the mobile robot, the calculation will be done on remote server and communicate with the robot through Wi-Fi connection. By this, the computation load on mobile robot can be reduced as well as improving the efficiency. A real-time feedback system is also built to promote accuracy in actual environment. After the development of improved stability and flexibility, the robot can be brought to real life as an interactive and collaborative robotic system.

Keywords: Interactive collaborative robotics · Mobile robot · Wireless charging · Probabilistic roadmap

1 Introduction

Nowadays, people rely more and more on mobile device such as smartphones, smartwatches, tablets and etc. Power transmission becomes a significant issue while more energy is needed for individual mobile devices user. We can see people try to find plugs for their devices in public. Instead of letting people to find the plugs, it is much more convenient for a robot to bring the energy to the user. By this, we develop a wireless charging robot which carried power bank and equipped with Qi charging transmitter use for charge devices support Qi charging.

The main idea of our research is to combine two existing technologies of wireless charging and mobile robotics in order to have optimum charging performance in static and dynamic environment for single or multiple nodes according to the state of the art in robotics and wireless charging. The robot could use probabilistic roadmap which solves the problem of determining a path between a starting point of the robot and a goal while avoiding collisions. With the Travelling Salesman Problem, we can extend our system to multiple devices. Wireless charging robot gets the command from server and

© Springer International Publishing Switzerland 2016
A. Ronzhin et al. (Eds.): ICR 2016, LNAI 9812, pp. 55–62, 2016.
DOI: 10.1007/978-3-319-43955-6_8

completes the charging mission with considering to spend least time and consume minimum power. Although mobile wireless charging robot is highly flexibility, there are many restrictions to the hardware. In order to reduce to computation load of the system, we aim to make the trajectory planning and navigate the robot to arrive to the destination precisely via Wi-Fi connection. The main computation would be on the server and each mobile robotic node would follow the instruction for wireless charging. We not only use the camera to capture the environment but also as an indoor localization resource. In addition, the camera acts as a real-time feedback system and sends data to the server to reduce inaccuracy occur in real environment. After developing a working test bed prototype and improvements, such system can be used in various environments such as office, lab, campus, air-ports, and train stations.

2 Background

2.1 Collaborative Robot

The concept of collaborative robot is that a human and a robot interact with each other. In our case, the robot charges the mobile device by receiving the demand from human. Although the robot did not interact directly with human, it assists human to charge mobile device to reduce human effort for charging.

2.2 Wireless Power Transfer

Charging mobile device wirelessly is an advanced approach. Although this charging method has been introduced to the world for a long time, it just becomes the trend recently. It allows portable electronics to charge them without ever being plugged in ubiquitous power wire. Transferring power without physical connection between the source and appliance reduces the charging process for the user. The most important advantage of wireless charging is to allow mobile devices to be charged when the transmitter and receiver are closed and lose the constraint of a power cord. By not having to

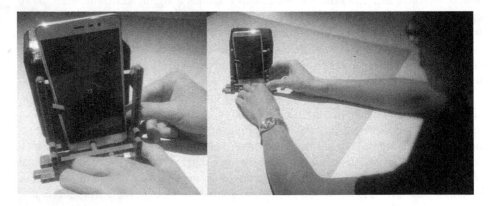

Fig. 1. Wireless charging robot and human interact with each other

deal with cords, the user doesn't have to worry about harming the wire by accident and the various types of chargers for various devices (Fig. 1).

The charging technology used in our case is Qi standard. This technology uses inductive coupling between two planar coils to transfer power from the power transmitter to the power receiver. In order to obtain enough magnetic fields, both transmitter and receiver need to be closed to each other. This is the reason why the robot needs to approach to the devices [1, 2].

3   Methodology

3.1   System Description

The case presented in this paper is to charge a mobile phone by wireless charging robot. For this purpose, a wireless charging robot was assembled with Arduino Uno, Arduino WIFI Shield, Go-between Shield by Mayhew Lab and Polulu Zumo robot for Arduino Ver.1.2 (Fig. 2). In order to make the system works, an ideal environment include a gray plane and a camera above it was set for the robot to accomplish this mission.

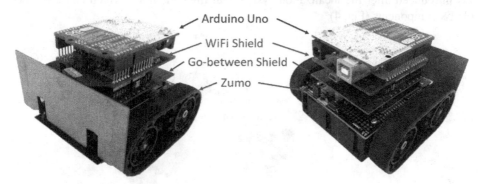

Fig. 2.   Inside view of the robot (without power bank and Qi transmitter)

On the gray plane which was set as the moving range for the robot, a few white blocks represent the barrier and black dots represent the mobile device need to be charged was placed. In the beginning, the camera captures a picture of the environment

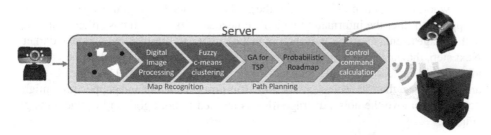

Fig. 3.   Flowchart of the system processing steps (Color figure online)

for the computer to build the map. The computer calculates the possible way for the robot and sends moving command to the robot through Wi-Fi. The red and blue rectangles on the robot are feature points for the computer to recognize the direction which the robot is facing (Fig. 3).

3.2 Robot Design

During the tests, we design different prototypes. Owing to the restriction of the localization system we designed, the robot should be whole black. In the beginning, we covered the robot with a simple black box which is light but not attractive. Furthermore, this design causes too much friction because it touches the floor and drag by the robot. Soon we finished our second design which improved the cons of first prototype. In this design, we considered the structure including the space for Qi transmitter and power bank. The bottom of the case is lifted from the floor so the friction problem was solved. During the test, a serious problem occurred. The design was too heavy for the Zumo robot to carry and the robot does not move as we expected. To solve the entire problem that founded during the tests, we modified the localization system. A simplified design was introduced after the localization system can filter the noise which cause by the robot's components (Fig. 4).

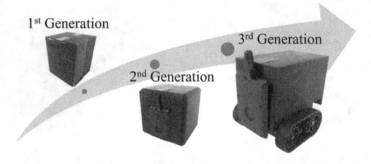

Fig. 4. Prototype evolution

3.3 Recognition of the Map

Digital Image Processing: In the beginning, the camera takes a bird's eye view picture of the whole environment. To recognize the goals and obstacles, digital image processing is applied to filter the information which we needed. First, a high pass filter is applied to the image to enhance the contrast ratio: make bright part brighter and dark part darker. Next step, the enhanced image is converting into gray scale which makes the value of all pixels become 0 to 255. By this, we set the value larger than 210 become obstacles and value lower than 30 become goals. After this process, the salt and pepper noise might occur. To remove the noise, average filter is applied to both goal and obstacle image (Fig. 5).

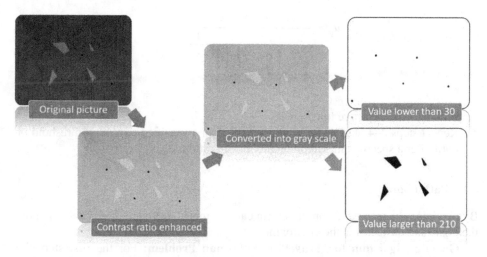

Fig. 5. Flowchart of digital image processing

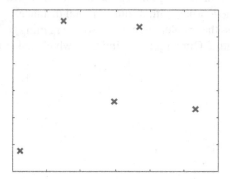

Fig. 6. Centers calculates by fcm function

Fuzzy c-means Clustering: After the image processing, fuzzy c-means clustering is applied to calculate the center of each goal. In this step, we applied fcm function in MATLAB. fcm performs the following steps during clustering (Fig. 6):

1. Randomly initialize the cluster membership values, μ_{ij}.
2. Calculate the cluster centers:

$$c_j = \frac{\sum_{i=1}^{D} \mu_{ij}^m x_i}{\sum_{i=1}^{D} \mu_{ij}^m}.$$ (1)

3. Update μ_{ij} according to the following:

$$\mu_{ij} = \cfrac{1}{\sum_{k=1}^{N} \left(\cfrac{\|x_i - c_j\|}{\|x_i - c_k\|} \right)^{\frac{1}{m-1}}}.$$ (2)

4. Calculate the objective function, J_m.
5. Repeat steps 2–4 until J_m improves by less than a specified minimum threshold or until after a specified maximum number of iterations [3, 4].

3.4 Path Planning

By obtaining the map which process form camera image, the computer is able to know the goals and obstacles in the environment.

Genetic Algorithm for Travelling Salesman Problem: For the first step, the computer decides the order of the goal by using Genetic Algorithm for Travelling Salesman Problem. The aim of using this technique is to move from one node to another exactly once and end the journey with minimum total distance.

For this case, express the problem in math is solve $\arg\min_x f(x)$, where x is the path. $f(x)$ is the distance of path x. Our target is to find the x which makes minimum (x) (Fig. 7).

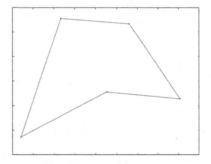

Fig. 7. Sequence decided by GA for TSP

The Genetic Algorithm stands on natural evolution rule, Survival of the fittest. It is a method which does not need to explore every possible solution in the feasible region to obtain a good result. Encoding, evaluation, crossover, mutation and decoding are the steps of GA process [5].

Probabilistic Roadmap: After decided the sequence of the goals, the computer calculate the way for the robot by Probabilistic Roadmap to avoid collision. In the beginning, random nodes are generated in the free space in the map (blue nodes in Fig. 8). Connections between each node are created and avoid crossing occupied space (gray lines in Fig. 8). The number of random nodes is carefully tuned after tests. Increasing the number of nodes can increase the efficiency of the path by giving more feasible paths. However, the increased complexity increases computation time. The additional nodes increase the complexity but yield more options to improve the path. The final path is created by finding an obstacle-free path using this network of connections with the shortest total distance [3, 6].

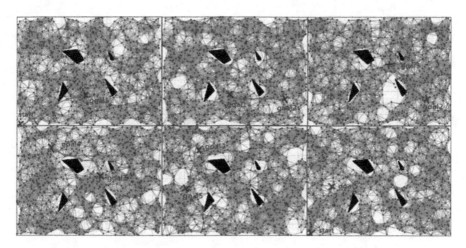

Fig. 8. Path decided by Probabilistic Roadmap (Color figure online)

3.5 Robot Control

In the final step, it is time to bring the simulation to reality. The Zumo robot is driven by two separate motors. The speed of the right motor is $v + \omega$ and the left is $v - \omega$. By adjusting the ω, we are able to change the direction which Zumo robot faced. The same method as calculating the centers of the goals was used to calculate the centers of the feature points which is use for knowing the location and the direction of the Zumo robot. Owing to the constrains of data type which the Wi-Fi shield supported, we need to encode the ω calculated by the computer and decode in Zumo robot (Fig. 9(a)).

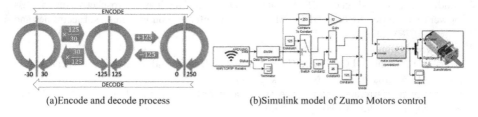

(a)Encode and decode process   (b)Simulink model of Zumo Motors control

Fig. 9. Control module of the mobile robot

4   Result and Conclusion

We have done many tests and adjust the variables to optimize the performance. We have measured and compared the execute time with different scenario including goal number and placement. According to the data acquired from tests, we found that the more concentrate the goals are, the less time is needed for path planning (Fig. 10).

Fig. 10. Wireless charging robot performance in an indoor environment

5 Future Works

To improve the charging performance, we are planning to extend our system to multiple robots. Furthermore, the localization system will be modified or changed to fit the environment which is similar to the reality. We believe that after modification, the wireless charging robot can be widely use in our smart society.

References

1. Shidujaman, M., Samani, H., Arif, M.: Wireless power transmission trends. In: International Conference on Informatics, Electronics and Vision (ICIEV), pp. 1–6 (2014)
2. Shidujaman, M., Rodriguez, L.T., Samani, H.: Design and navigation prospective for wireless power transmission robot. In: IEEE International Conference on Informatics, Electronics and Vision (ICIEV), pp. 1–6 (2015)
3. Mathworks online. http://www.mathworks.com/
4. Tanaka, K., Wang, H.O.: Fuzzy Control Systems Design and Analysis: A Linear Matrix Inequality Approach. Wiley, New York (2001)
5. Chatterjee, S., Carrera, C., Lynch, L.A.: Genetic algorithms and traveling salesman problems. Eur. J. Oper. Res. **93**(3), 490–510 (1996)
6. Amato, N.M., Bayazit, O.B., Dale, L.K., Jones, C., Vallejo, D.: OBPRM: an obstacle-based PRM for 3D workspace. In: ROBOTICS: The Algorithmic Perspective, pp. 155–168 (2002)

Distributed Information System for Collaborative Robots and IoT Devices

Siarhei Herasiuta[✉], Uladzislau Sychou[✉], and Ryhor Prakapovich[✉]

United Institute of Informatics Problems of the NAS of Belarus, Minsk, Belarus
{contacts,vsychyov,rprakapovich}@robotics.by

Abstract. In this paper we describe our experiments with a group of collaborative robots, IoT devices connected to robots based on WebSockets RFC 6455 Protocol using Google protocol buffers for message handling, distributed information system for it and a human interaction problem with these systems.

Keywords: Internet of Things · Robotics as a service · WebSockets · Collaborative robots · Google protocol buffers · Robo-hockey competition · Distributed information system

1 Introduction

In the IoT age, producing almost any machine going forward means giving it an IP address on the Internet. With robots we no longer just retrieve information from somewhere remote, we no longer just become tele-present somewhere else, we can now act on the physical world in a location other than where our physical body exists. This aspect of robotics is rarely discussed outside of a few existing examples such as military drones or healthcare. Yes, there are and will be many cases where fully autonomous robots will be designed to perform particular tasks for us. It's not necessary for us to be involved in controlling the operation of our vacuum cleaning robots. However, there will be just as many cases (and perhaps many more) where we will want to act, in a general sense, in a remote location, and where we'll share decision making and control of the robot, with the robot.

In Fig. 1, we demonstrate a very low size and low power new IoT device in comparison with "old" hardware, but they can control regular human oriented devices (refrigerators, washing machines and etc.) and communicate with robots. So there we see a very quick technology revolution for such type of devices.

2 "Robots as a Service" Approach in Building Distributed Information Systems

The basic idea is that run web service infrastructure support on a robot control system board. One of the solution is to use Web Sockets technology [1]. The WebSocket Protocol enables two-way communication between a client running untrusted code in a controlled environment to a remote host that has opted-in to communications from that

© Springer International Publishing Switzerland 2016
A. Ronzhin et al. (Eds.): ICR 2016, LNAI 9812, pp. 63–68, 2016.
DOI: 10.1007/978-3-319-43955-6_9

Fig. 1. Completely new IoT hardware ESP8266 ESP-05 with antenna (green small square) and old IoT hardware WiFly RN-XV 171 (red big square) placed in low level robot control system (Color figure online)

code. The security model used for this is the origin-based security model commonly used by web browsers. The protocol consists of an opening handshake followed by basic message framing, layered over TCP. The goal of this technology is to provide a mechanism for browser-based applications that need two-way communication with servers that does not rely on opening multiple HTTP connections (e.g., using XMLHttpRequest or <iframe>s and long polling).

Other idea is to decrease maximum data exchange between robots using well optimizing binary data protocol for objects handling over network channels. And the best solution there is using Google Protocol buffers. Google Protocol buffers are Google's language-neutral, platform-neutral, extensible mechanism for serializing structured data – think XML, but smaller, faster, and simpler. We can define how we want our data to be structured once, then we can use special generated source code to easily write and read our structured data to and from a variety of data streams and using a variety of languages.

3   Robot to Robot and Robot to Human Communication Problem

Social science literature indicates many types of human communication behavior used during collaborative tasks, including attentional cues to indicate an area of focus, staging actions to maximize shared visual information, gestural and speech cues indicating intentional goals or instructions, and coaching actions such as feedback, encouragement, and empathetic displays to build team rapport. Effectively producing all of these

Fig. 2. Group of different types mobile robots controlled by Android application using Wi-Fi protocol

communication actions on robots, in real-world task environments, is not currently feasible and would be difficult to generalize across different robot embodiments; for example, indicating attentional focus is different with humanoid and non-humanoid robots. Based on a review of the relevant social science literature covering human-human collaborations, and on observations of person-person task collaboration in our experimental setting, we focus in this work on using speech. Speech works well across different embodiments and is effective in communicating intent. On the other hand, speech has obvious limitations in noisy environments, with users with hearing or linguistic limitations, and in certain scenarios, such as disambiguating many similar objects. Nonetheless, speech is a natural human communication modality that addresses a range of use cases in home and work environments. Other methods of group robot control is not suitable if robots are different (Fig. 2).

4   Language Modelling for Robots-Human Interaction Using NooJ Approach Experiment

The goal of this experiment [3] is to interact with some number of robots in order to make them perform commands. With NooJ approach [2] this model can be designed in much easier way compared to other tools. The idea is to design the language that would be common and close to every-day language of the humans and that it would be able for machines to 'understand' it. Further design will be dedicated to replaying the model that has already been designed and the new data which is the new possible language

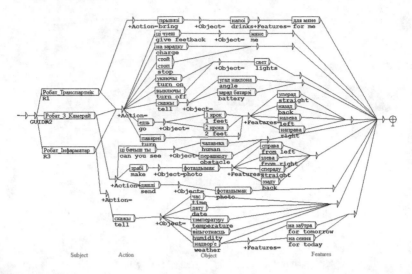

Fig. 3. Ontology semantic based high level standard of control system for three robots

constructions, phrases, linguistic units, etc. that can be expected from humans in order to interact with machines in their natural way.

At the start stage of the work we use deep syntactic analysis to get the model that is as simple as possible and yet far from underfitting the real model. We will use such concepts as 'Subject', 'Action', 'Object' and 'Features'. Using NooJ Syntactic Grammar we design graph model for combining all these concepts and linguistic units that will refer to them. Eventually we perform play-out routine to generate dictionary for robots using NooJ Dictionary. Some units from this dictionary for Belarusian language will look like:

> *Робат _ Інфарматар скажы надвор'е на заўтра,GUID=R3+Action=tell+Object=weather+Features=for tomorrow*
> *Робат _ 3 _ Камерай зрабі фотаздымак злева,GUID=R2+Action=make+Object=photo+Features=from left*
> *Робат _ Транспартнік прывязі напоі для мяне,GUID=R1+Action=bring+Object=drinks+Features=for me*

'Subject' (GUID in our example) refers to a robot's name. 'Action' refers to an action to be performed by robots that is usually represented by a verb. 'Object' represents a target of the action. And 'Features' is an add-on to specify 'Object' or 'Action'.

Using such kind of concepts, which are natural for humans, and NooJ tools we can generate Language Model for robots-human interaction and it will be high level stand of robot control system (Fig. 3).

5   Experiment of Controlling Two Groups of Robots Functioning Opposing Each Other

Experimental studies were carried out in two stages (Fig. 4). In the first phase conducted field tests using the experimental group of mobile robots. The group consists of eight robots and the central computer. Each robot is equipped with a wireless modem XBee Pro, working in the broadcast mode, the microcontroller the low-level control system and tracked chassis with dimensions $150 \times 190 \times 100$ mm.

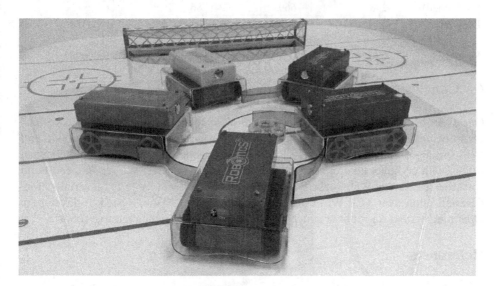

Fig. 4. Hockey with robots competition

Low-level control system implements a targeted reception of data packets from the central computer, and execute control commands, and send to the central computer of sensory data.

The central computer is equipped with a radio modem of the same type as the robots. Its software allows you to communicate with robots, data storage and the formation of management teams with interviews scrap-hundred milliseconds.

For the interaction of robots and the central computer network protocol is designed based on the physical and data link layers XBee Pro and implements the application layer of the OSI model, as well as elements of the network and transport levels to the extent necessary to control a group of robots.

Using existing robots possible to determine the basic relationships that characterize the information transfer processes in a group of robots. The dependence of the time the data packet on the packet size, the effect of the removal robot the signal strength (RSS) and the related transmission errors, the maximum number of simultaneously operating the transmission in the broadcast mode radio modem [4–7].

Table 1. Sterilisation/deserialization performance comparison for small and large objects in different common used data object formats (Size in bytes, time in milliseconds)

	XML	Binary	JSON-service stack	Protocol buffer
Size (large)	323.981	204.793	141.863	104.191
Deserialize (large)	7.889384	19.39763	5.731472	3.82069
Serialize (large)	5.508091	13.700064	3.559688	1.447036
Size (small)	298	669	86	62
Deserialize (small)	0.015977	0.019405	0.00174	0.003883
Serialize (small)	0.021897	0.021023	0.003645	0.000989

6 Conclusions

Our experiments give us vision that we need to create a new programming algorithms for software development of distributed information system for collaborative robots and IoT. This algorithms should adopt well known approaches to a new class of Intellectual robots with complex control system and IoT enabled devices. We recommend build software stack using Web Sockets, Google Protocol Buffers (Table 1) for serialization/deserialization over TCP Wi-Fi high speed connection. Also we need to develop an Operating system for robot or adopt current used (ROS) for collaborative robots.

References

1. Fette, I., Melnikov, A.: The WebSocket Protocol RFC 6455. Internet Engineering Task Force (2011)
2. Silberztein, M.: Formalizing Natural Languages: The NooJ Approach. Wiley, Hoboken (2016)
3. Kaigorodova, L.: Language modelling for robots-human interaction. In: International Scientific Conference on the Automatic Processing of Natural-Language Electronic Texts "NooJ 2015", UIIP NASB, Minsk, pp. 28–29 (2015)
4. Herasiuta, S., Sychou, U., Prakapovich, R.: Development of protocol, hardware and software platform for centralize robot group control. In: Automation and Robotic Application for Processes and Manufacturing, Businessofset, Minsk, pp. 101–102 (2014) (in Russian)
5. Herasiuta, S., Sychou, U., Prakapovich, R.: Size scaling features investigation for group of small-sized mobile robots based on centralized and decentralized method of managing. In: Extreme Robotics, Saint-Petersburg, pp. 76–81 (2015) (in Russian)
6. Herasiuta, S., Sychou, U., Prakapovich, R.: Development of open technology designing of intelligent robotic systems. In: Open Semantic Technologies for Intelligent Systems, BSUIR, Minsk pp. 487–488 (2015) (in Russian)
7. Herasiuta, S., Sychou, U., Prakapovich, R.: An experiment for mapping with group of robots. In: Eight All-Russian Multi-conference for Control Problems, pp. 161–164. Sought Federal University Publishing, Gelendzhik (2015) (in Russian)

Designing Simulation Model of Humanoid Robot to Study Servo Control System

Alexander Denisov$^{(\boxtimes)}$, Victor Budkov, and Daniil Mikhalchenko

SPIIRAS, 39, 14th Line, St. Petersburg 199178, Russia
sdenisov93@mail.ru

Abstract. In the paper, we consider a method of developing a simulation model of a humanoid robot preserving its size and mass-inertial characteristics. We describe a developed model of the servomotor and a method of tuning its PID controller as well as a method of robot interaction with the surface based on the PD controller. With the help of a special method of tuning the PID controller in Simulink program, controls coefficients for the servomotor models were selected. The experimental results have shown that the servomotors can reach a predetermined angle of rotation and maintain this position with sufficient accuracy. We have conducted a simulation of the interaction of the humanoid robot with the floor surface. With Simulink software blocks, we have made a switch that activates support reaction force at the transition of the robot foot coordinates along a vertical axis to a negative value, which provides an adequate interaction of the model of the robot with the floor model.

Keywords: Humanoid robot · Mass-inertial characteristics · Servomotor · PID controller

1 Introduction

Currently, modeling is one of the most important and indispensable stages of designing mechatronic devices and robotic systems. To make the development process faster, as well as identify and eliminate errors at an early stage, various simulation software tools are used. The feasibility of using simulation to design robots is defined by the following factors:

- Provision of building models with a large number of elements of the system under development;
- Carrying out complex experiments that are not available for a real object;
- Visibility of the experiments and results;
- Provision of monitoring of changes in the values of variables over time;
- Solution of tasks that are impossible to solve analytically;
- Determination of the most important variables of the system and their interactions.

In [1], a method of modeling and simulation of the robot using SimMechanics library is presented. The physical modeling, parameter setting, and simulation of the model are considered in detail. The paper describes the creation of the model of the robot in SimMechanics to carry out experiments on the robot gait using a method of

© Springer International Publishing Switzerland 2016
A. Ronzhin et al. (Eds.): ICR 2016, LNAI 9812, pp. 69–78, 2016.
DOI: 10.1007/978-3-319-43955-6_10

inverse pendulum as an example of the classical nonlinear control experiment. A mathematical model of the pendulum is built taking into account the equations of Newtonian dynamics, the model itself is made from SimMechanics blocks. The development of a feedback system to meet the imaging requirements is described, which enabled the authors of the paper to make a conclusion about the possibility of using SimMechanics for unstable nonlinear systems and robots. The authors make the following conclusions:

- There is no need to calculate the direct problem of dynamics and output complex differential equations, in which it is easy to make a mistake, for modeling in SimMechanics;
- Model, built in SimMechanics, can be used to test the control system.

In [2], the development of a humanoid robot with the help of Virtual Reality Modeling Language (VRML) and the simulation of the model in SimMechanics is considered. Since a robot production is a costly process, a simulation model has been developed to conduct experiments on the robot in the absence of the possibility of real experiments. The model of a humanoid robot allows one to study the kinematics, dynamics and a control method of a real construction before production of a prototype. Technology for creating a robot model, discussed in this article, allows one to simulate the entire object, rather than individual joints, so changing the size of one part will lead to the automatic change of the size of the entire structure in order to preserve relations between the parts. This article presents a simple, reliable, cheap and easy-to-use way to develop robots in SimMechanics and VRML. In addition, this model allows one to carry out experiments on the control systems of gait and work of robot arms.

The paper [3] considers the development of the geometric and mathematical model of a robotic arm Mitsubishi RV-2AJ with 5 degrees of freedom. The model of the robot is designed using SolidWorks CAD software and SimMechanics library in Simulink. The article describes the process of transferring the model from CAD to Simulink/ SimMechanics environment. With the preservation of the models of manipulator parts as XML files, the size and mass-inertial characteristics of the construction do not change when they are further connected to SimMechanics blocks. The authors consider the development of a feedback control system, the results of which show that this model is 98.99 % the same as Mitsubishi's actual manipulator.

In [4], the authors consider the development of the model of a humanoid robot in Simulink /SimMechanics environment as well as mobility algorithms that enable the robot to achieve a stable anthropomorphic gait. An experimental prototype of the Russian human-sized robot - a robot AR-601 M- is used as a simulation object. The model of the robot, unlike the original, has 11 degrees of freedom, since at this stage it is intended only for modeling the robot's gait. The gait of the model is based on the concept of ZMP (zero moment point) and the inverse pendulum. For a dynamically stable gait the model is tracking the initial position of the robot. During the simulation, the maximum speed for a stable gait of the model is 0.3–0.4 m/s. The authors note that this is only the initial development. Further development of the project will be aimed at increasing the degrees of freedom by adding new joints in the model for full compliance with the robot AR-601 M as well as at a more detailed development of humanoid gait using MotionCapture system (MoCap).

Taking into account the recent research devoted to design of anthropomorphic robot construction the own simulation model was investigated. Further, the developed simulation model of an anthropomorphic robot for carrying out experiments on the design, tuning of servomotors and developing a control system is described.

2 Development of Simulation Model of Humanoid Robot

To solve the problem of development of the humanoid robot model, a Simulink program of MATLAB software package is used. Simulink is a graphical simulation environment that allows one to build dynamic models, including discrete, continuous and hybrid, nonlinear and discontinuous systems using block diagrams in the form of directed graphs. Simulink combines the clarity of analog machines and the accuracy of digital computers, possessing all the capabilities of MATLAB.

Simulink implements the principle of visual programming whereby a user creates a model of the object under study, using building blocks of the library. In the interactive simulation environment it is possible to change parameters even at runtime of model, thereby observing alterations in the behavior of the object and results of its operation in real-time.

SimMechanics Library is a separate library of Simulink/MATLAB software, intended for the simulation of the mechanical movement of solid objects, according to the laws of theoretical mechanics [5]. Similar to all the models developed in Simulink, the model in SimMechanics is represented as a block diagram of the relevant building library blocks. Mechanical blocks of SimMechanics represent physical objects or connections between them. Using the library one can model a mechanical system consisting of any amount of solid objects. Joints with translational and rotational degrees of freedom are used to connect solid objects. Assigned parameters of SimMechanics block are the mass-inertial properties of objects, coordinates of the main points such as the centers of mass, connection points of joints, points of application of external and control actions. Kinematic constraints, forces and torques can be applied in blocks [6]. SimMechanics allows measuring motion parameters in the process of modeling.

SimMechanics uses MATLAB visualization tools for spatial representation of the mechanical system throughout the modeling process. In addition to the simplified display in the form of approximating polygons or ellipsoids, the expansion package of virtual reality VIRTUAL REALITY TOOLBOX can be used, which allows to link the model of the solid object with the appropriate file and ensures monitoring of the movement of the model during the simulation in the VRML browser.

Building a model of the mechanism in SimMechanics includes the following steps:

- Determination of the mass-inertial characteristics of solid objects, their degrees of freedom and constraints in accordance with the objects coordinate systems;
- Inclusion in the model of virtual sensors for measuring parameters and virtual motors for ensuring movement of mechanisms, attached forces and moments;

- Launching the simulation process (implemented on Simulink platform) and the study of the movement of the mechanism until it stops because of existing restrictions;
- Visualization of the simulation process (mechanism movement) during the overall simulation process in a special graphic window [7].

Figure 1 presents a model of the robot under development. Dimensions of the model of the robot are the following: height is 906 mm; width of the shoulders is 317 mm; width of the foot is 170 mm; weight is 8 kg. This model is a prototype, and its design is being improved, so the size, weight and type of the model change in the course of development.

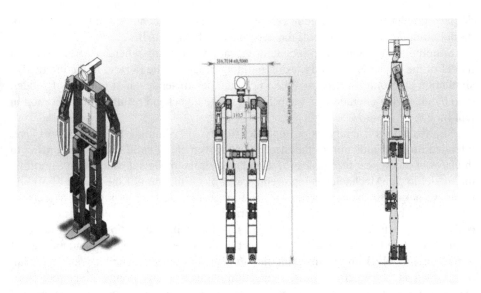

Fig. 1. A model of the robot being developed in SolidWorks CAD software

The model consists of 21 blocks of solid objects of Body library in SimMechanics, representing robot design elements: feet, lower legs, knees, etc. These blocks are connected via joints performing rotational degree of freedom along one of the axes Ox, Oy, Oz. Each Joint block is connected to the position sensor and a servomotor model which receives a control signal specifying the angle of rotation of joint. The model of the robot has 22 rotational degrees of freedom, which ensures spatial motion of the robot, movement of its arms, a tilt and turn of its head to assess the environmental by means of a robot camera. The model was made in the SolidWorks CAD software, where mass-inertial characteristics of the design elements were calculated, and transferred to SimMechanics environment with the help of the utility simmechanicslink preserving all calculated characteristics.

3 Tuning Parameters of the Model of Servomotor of Humanoid Robot

To move the model in a virtual space and to perform any other movements, actuators must be connected to the model hinges. In this construction, servo models with a DC motor are used as actuators. This type of motor has been selected due to its sufficient capacity and small size. Servo model in Simulink is shown in Fig. 2.

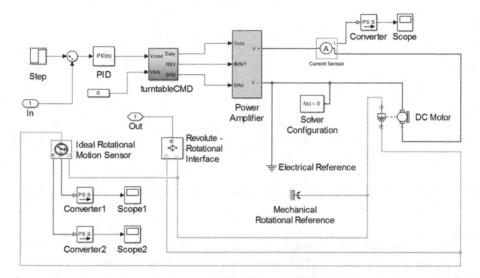

Fig. 2. A model of servomotor in Simulink

Servomotor model consists of blocks of SimElectronics library - electronic component library. Figure 2 shows that the servo model consists of a control signal generator (block Step); the adder that receives the negative value from motor rotation sensor, which forms a feedback to the motor; a proportional–integral–derivative controller (PID controller); a block of DC motor. In the PWM Rate subsystem there is a system for calculating the duty cycle; voltage value at which the motor will run in the opposite direction; a short circuit unit. In the subsystem Power Amplifier there are PWM and H-Bridge blocks. In the right part of the scheme there are the DC motor block and sensors connected to it, a special block for connecting the motor model to the elements of the library of SimMechanics - Revolute-Rotational Interface. DC motor block represents electrical and rotating DC motor characteristics, using the model of electric circuit consisting of power supply, inductor and resistance [8].

PID controller of the model must be tuned for accurate rotation through a predetermined angle. For the most efficient operation of the PID controller it is necessary to select gains for each controlled object individually. In Simulink there is a specialized block that represents the given controller, whose coefficients can be set manually or automatically. For automatic selection of coefficients PID Tuner subroutine is used.

Figure 3 shows graphs of the transition process of servomotors operation at assigning a specific angle of rotation of the first knee and hip joints. Figure 3.a shows the transient characteristics of the joint rotation at robot's knee through 30°; Fig. 3.b - rotation in the hip joint of the model through 60°.

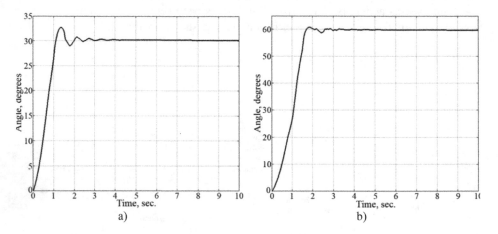

Fig. 3. Servomotors operation: (a) transient characteristics of the knee joint, (b) transient characteristics of the hip joint

These graphs show the transient characteristics of the rotation angles of servomotors; the overshoot of the PID controllers; an ability of servomotors to hold a predetermined angle. The obtained transient characteristics allow us to judge about the correctness of tuning the controller. As can be seen from Fig. 3, the actuators reach the predetermined rotation angle and maintain this position with sufficient accuracy. The first graph shows the overshoot by 2.5°, but in this design construction such a small overshoot can be neglected. It follows from the graphs that the automatic adjustment of PID controller coefficients is successful, and the motors are working correctly.

4   Experimental Verification of the Robot Simulation Model

Using the building library blocks SimMechanics it is easy to simulate complex systems consisting of many objects and joints, but for this, the model needs to have a rigid connection with the fixed surface (support), as the library does not have the automatic recording of interactions between objects. To realize this interaction, we developed a simplified surface model. In this case, it was a model of the floor as the object in which with the presence of the load (forces acting on the surface) there occur forces that oppose the load, thereby preventing the deformation of the object. A model was constructed on the basis of Simulink blocks (Fig. 4).

Figure 4 shows the scheme of interaction of the model of the robot with fixed support, built similarly to a proportional derivative (PD) controller. Block Body Sensor,

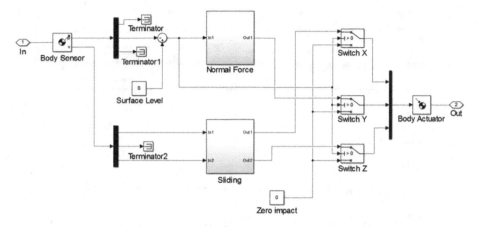

Fig. 4. A scheme of robot interaction with the floor

determining a change of coordinates and velocity, is connected to the robot foot. Block Surface Level sets the height wise shift of level of the surface in the global coordinate system. Blocks Switch X, Y, Z are the control blocks of the support reaction force along the Y-axis and of sliding along the axes X and Z. If there is a positive change in the robot foot coordinates along the vertical Y-axis relative to the value in block Surface Level, in this case, Y > 0, then block Switch Y receives the signal from Zero Impact block. This models the absence of reaction from the surface, since the foot is located above it. If the coordinate of the Y-axis becomes smaller than a predetermined value in block Surface Level, then the block Switch Y takes values of the subsystem Normal Force equal to the support reaction force. The floor is modeled as a spring which has a force opposing the load, so for the subsystem Normal Force coefficients of elasticity and damping are used. The subsystem Sliding presents the blocks for calculating sliding on the surface along the axes X and Z respectively. In this case, the value of blocks is equal to zero, which simulates the absence of sliding on the floor surface. Values from block Switch arrive at the block Body Actuator, which generates a corresponding force acting on the object and preventing it from falling through the surface.

Figure 5 shows a diagram of the mass center of the foot of the robot being developed. There is practically no change in position of the mass center of the foot along the Y-axis (upper line); changes occur during the first 2 s of simulation and are equal to about 1 mm, which is associated with the selected elasticity and damping coefficients. The position of the center of mass along the X-axis (middle line) does not change and is equal to zero as under the initial conditions.

Along the Z-axis (bottom line), there is a slight sliding in the positive direction approximately equal to 1 mm for the first 2 s of the simulation, which satisfies error conditions of the model being designed. As is clear from Fig. 5, the model of the surface allows the object to stand on it in view of the surface elasticity and sliding friction.

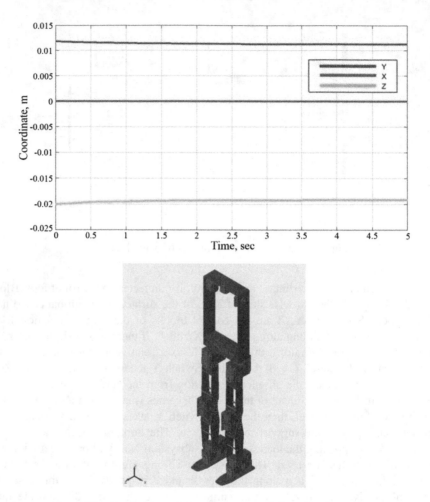

Fig. 5. Position of the mass center of the robot foot

The initial works at design of anthropomorphic robot was started for investigation legs [9]. A twin-engine layout, used in the knee joint, ensures higher joint power along with independent interaction with the neighboring hip joints and tibia joints when bending. The larger bending angle of knee joint relative to the single-engine layout is achieved. The number of degrees of freedom of each leg integrated with pelvic mechanism rises to 7. At the stage of the prototype manufacture, the Dynamixel actuators produced by Robotis are used to validate and debug the units operation. To reduce the electrical load on the main battery of the robot, the femoral parts of the legs are provided with a mounting pad for additional batteries powering servos. The application of the anthropomorphic robot is focused on the educational purposes, particularly on participation in robot soccer competitions, as well as on the development of assistive technology of human-computer interaction based on multimodal interfaces [10–15].

5   Conclusion

With the use of Simulink software, a model of the anthropomorphic robot with ser-vomotors, connected to joints, was developed; PID controllers were tuned; the system of interaction with the floor was designed. The most suitable servomotors for the real robot can be chosen taking into account mass-inertial characteristics of parts and due to the type of actuator. The model has the same number of degrees of freedom as a real robot being developed as well as the same mobility of joints. Thus, the constructed model allows one to conduct experiments on the robot control system at the stage of design engineering, which reduces the creation time and cost of the robot. A disad-vantage of the given model and of using simulation modeling is that there are elements in the model, some components of which are approximate. Thus, the model does not fully correspond to the real object and only allows estimating the real robot and its behavior to a certain degree. In the future, we are planning to improve models in order to correspond more accurately to real objects and connect them to the motion control system as well as to simulate movement of the robot in space.

Acknowledgment. The study was performed through the grant of the Russian Science Foun-dation (project №16-19-00044).

References

1. Shaoqiang, Y., Zhong, L., Xingshan, L.: Modeling and simulation of robot based on Matlab/SimMechanics. In: 27th Chinese Control Conference, pp. 161–165. IEEE (2008)
2. Zheng-wen, L., Guo-liang, Z., Wei-ping, Z., Bin, J.: A simulation platform design of humanoid robot based on simmechanics and VRML. Procedia Eng. **15**, 215–219 (2011)
3. Ayob, M.A., Zakaria, W.N.W., Jalani, J., Tomari, M.R.M.: Modeling and simulation of a 5-axis RV-2AJ Robot using SimMechanics. Jurnal Teknologi **76**(4), 59–63 (2015)
4. Khusainov, R., Shimchik, I., Afanasyev, I., Magid, E.: Toward a human-like locomotion: Modelling dynamically stable locomotion of an anthropomorphic robot in simulink environment. In: 12th International Conference on Informatics in Control, Automation and Robotics (ICINCO), vol. 2, pp. 141–148. IEEE (2015)
5. Lazarev, Yu.F.: Modelling of processes and systems in MATLAB: Training course. Piter, St. Petersburg, 512 p (2005)
6. Shcherbakov, V.S., Korytov, M.S., Ruppel, A.A., Glushets, V.A., Milyushenko, S.A.: Modeling and visualization of motion of mechanical systems in MATLAB: Training Course. SibADI Publ., Omsk (2007)
7. Tikhonov, K.M., Tishkov, V.V.: SimMechanics Matlab as means of modeling the dynamics of complex robotic systems. Electron. J. Tr. MAI, 41, 6
8. SimMechanics^{TM}. User's Guide: Version 2. Textbook, 522 p. The Math Works, Inc., Natick (2002)
9. Yusupov, R.M., Ronzhin, A.L.: From smart devices to smart space. Herald Russ. Acad. Sci. **80**(1), 45–51 (2010). MAIK Nauka
10. Karpov, A.A., Ronzhin, A.L.: Information enquiry kiosk with multimodal user interface. Pattern Recogn. Image Anal. **19**(3), 546–558 (2009)

11. Ronzhin, A.L., Budkov, V.Yu.: Multimodal interaction with intelligent meeting room facilities from inside and outside. In: Balandin, S., Moltchanov, D., Koucheryavy, Y. (eds.) ruSMART 2009. LNCS, vol. 5764, pp. 77–88. Springer, Heidelberg (2009)
12. Ronzhin, Al.L, Budkov, V.Yu., Ronzhin, An.L: User profile forming based on audiovisual situation analysis in smart meeting room. SPIIRAS Proc. **23**, 482–494 (2012)
13. Ronzhin, A.L., Karpov, A.A., Leontyeva, An.B, Kostuchenko, B.E.: The development of the multimodal information kiosk. SPIIRAS Proc **5**, 227–245 (2007)
14. Motienko, A.I., Makeev, S.M., Basov, O.O.: Analysis and modeling of position choice process for transportation of the sufferer on the basis of bayesian belief networks. SPIIRAS Proc. **43**, 135–155 (2015)
15. Pavlyuk, N.A., Budkov, V.Yu., Bizin, M.M., Ronzhin, A.L.: Design engineering of a leg joint of the anthropomorphic robot antares based on a twin-engine knee. Izvestiya SFedU Eng. Sci. **1**(174), 227–239 (2016)

Educational Marine Robotics in SMTU

Igor Kozhemyakin, Kirill Rozhdestvensky, Vladimir Ryzhov,
Nikolay Semenov, and Mikhail Chemodanov[✉]

Saint Petersburg State Marine Technical University, St. Petersburg, Russia
{kozhemyakin,kvr,ryzhov,semenov,chemodanov}@smtu.ru

Abstract. The concept, general provisions and some results of the research &
education initiative of Saint-Petersburg State Marine Technical University
(SMTU) «Marine Robotics: Education through Research» (MRER) are considered in the paper. The directions of research conducted at SMTU in the field
of marine robotics, as well as up-to-date technologies of training of future
engineers specializing in a certain area are represented. Some work results of
a student team, developing a small unmanned remotely operated vehicle (ROV)
are presented to exemplify the «through project study» approach. One of the
innovative solutions validated on this ROV is implementation of hydroacoustic modem based on polyphase filters and employed for positioning and
communication with other AUV.

Keywords: Education marine robotics · Education through · Research approach ·
ROV · AUV · Underwater glider · Wave glider

1 Introduction

SMTU has considerable experience in both investigations, aiming at the development of
modern marine technical objects and systems, and in application of advanced technologies, supporting educational process. Combining these two types of activities enables to
teach students of all three educational levels with much higher efficiency. This approach
promotes the development of creative professional competencies of the future engineer,
specializing in applied marine sciences. The SMTU MRER initiative under realization at
present time correlates eMaris project [1] (TEMPUS program), executed earlier, and with
the just submitted InMotion project [2] (program Erasmus +). Use of the Education
through Research approach gives positive practical results, associated with both enhancement of quality of professional training of students and with creation of innovative developments, carried out with direct participation of students, and attractive for potential
customers. In this connection special attention is allocated to highly demanded direction –
marine robotics, which comprises solutions of complex multidisciplinary problems
providing creation of novel prototypes of marine techniques, possessing unique operational properties. These tasks are being solved with participation of specialized academic
departments and laboratories, carrying out their activities in close contact with the contin-
gent of students.

© Springer International Publishing Switzerland 2016
A. Ronzhin et al. (Eds.): ICR 2016, LNAI 9812, pp. 79–88, 2016.
DOI: 10.1007/978-3-319-43955-6_11

2   Educational Robotics at SMTU

Use of the Education through Research approach implies acquisition by the students of miscellaneous knowledge in the process of conducting concrete scientific research projects as members of design teams.

Supported in the frame of this approach are:

- collective and individual design studies of students;
- use of up-to-date classes and development test-beds, equipped with simulation (imitation) systems for virtual elaboration of technical solutions;
- support of the «through project learning» by industrial enterprises by means of online lectures and consultations in the course of realization of the projects (with use of videoconferencing systems of high definition);
- participation of students with their projects in the international competitions.

Students of different academic specializations can take part in a wide spectrum of research activities in research directions related to: hydrodynamics, dynamics of objects, structural mechanics, designing of the object and its systems, information & measurement complexes, communication systems, power plants, artificial intelligence, automatic control systems, mission planning.

In its research and educational projects the SMTU is orientated toward prospective directions of the development of marine robotics, including:

- development of advanced platforms for underwater and surface applications;
- development of new propulsors, drives and power plants;
- development of efficient systems for navigation and positioning under water (IMU, correlation lags, USBL, LBL);
- development of systems for control, processing of the information from gauges, sonars and optical cameras, enabling to solve difficult problems autonomously;
- furthering of cooperative use of marine robotic systems.

3   Research and Educational Projects of SMTU Within the MRER Initiative

At SMTU there has been carried out a number of research and educational projects both of complex, and also related to the development of separate components of underwater robots. Among such projects are (see Fig. 1):

- the project of underwater vehicle «Aphalina» with bionic propulsor [3];
- development of flapping fin propulsor [4],
- the project of small-sized inspection ROV [5];
- the project of underwater glider [6];
- scaled mock-up of wave glider [7];
- development of an underwater communication system [8];
- development of compact sectoral search sonar system.

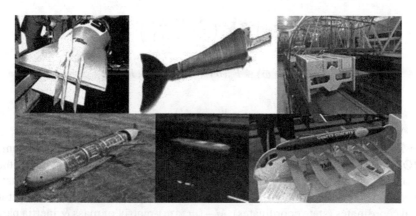

Fig. 1. Underawter vehicle «Aphalina» with flapping wing propulsor [3], bionic propulsor of underwater robot [4], small-sized inspection ROV [5], SMTU underwater glider [6], SMTU scaled mock-up of wave glider [7], SamGTU-SMTU wave glider (underwater module)

In the process of the development of objects and systems widely used are open platforms both software and hardware. Open software examples include OpenFoam, ROS, Gazebo, UWSim, and Octave. Open hardware platforms can be exemplified by microcontrollers Arduino, single-board microcomputers RaspberyPi, BeagleBone Black/Green, various drivers of collectorless electric motors and collectorless motors proper.

An exemplary education-through-research project is that for the development of an experimental ROV [5]. The goal of the project was in that the students should acquire experience in creation of a light-class ROV with a perspective of further development of the vehicle toward a partial and full automation as well as its use in inspection operations and for solution of some salvage tasks.

At the stage of the concept development, there were specified concrete technical requirements for this vehicle:

- speed of motion: in cruise - 2 m/s; lateral - 0.5 m/s; vertical: 0.25–0.5 m/s;
- maximum submersion depth up to 50 m;
- maximum electric power consumption - 3000 W;
- availability of a system of depth retention;
- availability of a system of automatic course retention;
- mass of the vehicle up to 15 kg, mass of the land-based (ship-based) equipment up to 10 kg;
- vehicle dimensions - 800 × 500 × 400 mm;
- securing possibility of work in currents with speeds up to 1 m/s.

Second stage – development of mathematical model of ROV, which enables evaluation of sea-going and maneuvering parameters of the vehicle under development. Mathematical model of underwater vehicle, based on the known motion equations of a solid body, can be represented in the following vector-matrix form [9]:

$$\dot{Y} = \sum (\bar{\theta}, \bar{X}) = \sum \left(\frac{\Sigma_P (\bar{\theta}, \bar{X})}{\Sigma_\theta (\bar{\theta}, \bar{X})} \right), \tag{1}$$

$$\tilde{M}\dot{\bar{X}} = [\bar{F}_d(\bar{P}, \bar{V}, \bar{\omega}) + \bar{F}_u(\bar{\delta}) + \bar{F}_v(G, A_\Pi, R_\Gamma)], \tag{2}$$

$$T_{uy}\frac{d\bar{\delta}}{dt} + \bar{\delta} = \bar{\Psi}_{uy}(\bar{\delta}, \bar{U}), \tag{3}$$

where T_{uy} and $\Psi_{uy}(\delta, U)$ – are correspondingly: diagonal matrix of time constants of the ROV hull and vector of nonlinear functions of right-hand side of the actuator equations; δ – is a vector of control action upon the ROV elements due to actuators; U – vector of controls, formed by the ROV control system, where X– m-vector of internal coordinates (state coordinates); M – $(m \times m)$-matrix of mass & inertia parameters; $F_u(x, Y, \delta, l)$ – m-vector of control forces and moments, here l – is a vector of construction parameters; $F_d(x, Y, l)$ – m-vector of nonlinear elements of the ROV dynamics; F_v – m-vector of measured and not measured external perturbations; $Y = (P, \theta)^T$ – n-vector of position P and orientation θ (output coordinates) of the body coordinate system relative to the basic coordinate system, $n \le 6$; $\Sigma(\theta, x)$ – n-vector of kinematic links; $\Sigma_P(\theta, x)$, $\Sigma_\theta(\theta, x)$ – vectors of linear and angular velocities of the body coordinate system relative to the basic coordinate system.

The coefficients of hydrodynamic forces and moments as well as the added masses of the ROV under design have been determined in [10]. Screw propellers were chosen as a propulsor-rudder complex of the ROV for the maneuverability at small speeds. In order to analyze maneuverability of the vehicle there was formalized a distribution of control forces and moments of a concrete configuration of propulsor-rudder complex with cruise ducted screw propeller and bow thruster (Fig. 2).

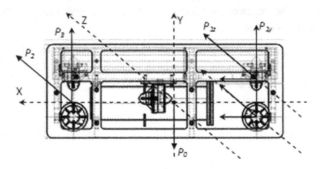

Fig. 2. Scheme of distribution of control forces of propulsor-rudder complex of the ROV (P_{1x}, P_{1y}, P_{1z} – projections of thrust forces, generated by the cruise ducted screw; P_2, P_3. – thrust forces, generated by the bow thrusters in the horizontal and vertical directions)

Chosen for the evaluation of controllability of the underwater vehicle was a standard approach, consisting in that: for the object to be controllable, it is sufficient that the thrust of the propulsion devices and hydrodynamic forces should exceed the required control forces along coordinate axes [9]. In particular, conditions of controllability can be written in the following form:

$$P_{1x}^{max} > |F_x|, P_{1y} + P_3 > |F_y|, P_{1z} + P_2 > |F_z|,$$
$$-P_{1z}x_{md} + P_2 x_{hz} > |M_y|, -P_{1y}x_{md} + P_3 x_{vr} > |M_z|,$$ (4)

where P^{max} – is a maximum thrust force, generated by the propulsor along the corresponding axis, F_x, F_y, F_z, M_y, M_z – projections of the vector of forces and moments upon the axes OX, OY, OZ.

As all control channels are used to control either thrust of the cruise propulsor or that of the thrusters, control of position and motion is available along all axes simultaneously.

Thus, the constructed mathematical model allows to take account of the hull form, position and parameters of the propulsors, and to calculate the ROV motion parameters.

Finally, making use of the developed mathematical model, the project participants have chosen an appropriate hull form, position of the propulsors, their power, position of the centers of gravity and buoyancy.

Knowledge of these characteristics allowed passing to design stage (elaboration of the structure of the vehicle) and software development.

The ROV was designed with use of the SolidWorks 2013 CAD system. There was created and elaborated a 3D model of the object, and released construction documentation necessary for manufacturing and assembly of the vehicle.

The ROV hull included three main parts: a supporting frame, elements providing neutral buoyancy (floats) and pressure hull (sealed enclosure for electronic equipment) (Fig. 3).

Fig. 3. The hull components: frame, element providing neutral buoyancy and pressure hull for the equipment

Details of the hull of the robot and fins were manufactured using milling machine with numerical control. For manufacturing details of the hull there were used polypropylene plates.

The pressure hull was an aluminum cylinder with a flange and a cover. Fixed to the external cover was a special internal casing comprising electronic equipment. On the cover there were placed sealed cable inlets for connections of the motors, gauges and power and control cables of the ROV.

Six collectorless motors T200 of BlueRobotics [11] were used as cruise and side motion propulsors.

Locations of propulsors of the ROV are schematically shown in Fig. 4. A variant of the propulsion installation configuration, adopted for the vehicle under design, envisages turning of the work sector of the vehicle (manipulator, camera, and lighting) rather than direct turning of the propulsors with the vehicle being constantly orientated against the current. Additional advantage of such a system is a possibility of auto-stabilization of the ROV against the current by means of using the keel.

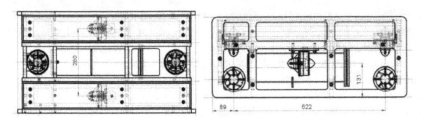

Fig. 4. Scheme of location of the ROV propulsors

The stage of the software development and on-board electronics was executed in parallel with the development of the vehicle hull.

Often times the ROV software is distributed i.e. operates simultaneously on several computers. In the system under discussion, the software operates on both shore and the underwater vehicle. In the chosen approach, the software is launched on two computers and is synchronized through single-rank local system.

Besides, the system implies use of standard periphery, including USB connected (joystick, camera, etc.).

The problems of use of the periphery and local network synchronization have been solved long ago. Therefore, to accelerate the development of software and to enhance reliability of the system it appeared reasonable to use ready and verified solutions. After analysis of different software libraries there was taken a decision to employ framework ROS (robot operating system) [12].

Structural scheme of the developed software and the source code of the system are accessible at the address [13].

Installed on the ROV are the gauges of depth, micromechanical accelerometers, gyroscopes and a magnetometer. In the course of the development of the project additional gauges can be connected with use of interfaces I2C, SPI, USB.

To elaborate the control system there was conducted an imitation modeling of the ROV motion in water with use of open software complexes of the type of UWSim, Gazebo. The goal of this work consisted in elaboration of the control algorithms, evaluation of their stability and reliability in different situations and their subsequent transfer into the developed ROV. In the course of the imitation modeling there were also found some deficiencies of the ROV structure, which later on have been eliminated. Eventually, the elaborated software was transferred on the ROV to get the vehicle prepared for the trials (Fig. 5).

Fig. 5. Real ROV and it's simulation model in Gazebo

For positioning of the ROV and communication between it and other underwater vehicles, the ROV is equipped with a hydroacoustic modem. Represented in the market at present time are hydroacoustic modems of the companies Evologics [14], Aquatec [15], Sonardyne [16], LinkQuest [17], Shtil [18] and others. These up-to-date modems are capable of transferring data either with speeds up to 48 Kbit/s at distances up to 1–2 km (modem S2CR 40/80 of EvoLogics GmbH), or up to 7–10 km with a speed 2,5 Kbit/s (UWM10000). Implemented in these developments are: phase manipulation, patented modulation technology S2C, frequency modulation and so on. Declared modulation rates are reached at certain conditions in hydroacoustic channel. However, real conditions are diversified and such phenomena as refraction, dissipation, reverberation result in distortions of the acoustical signal on receiving side. These distortions of the received signal significantly reduce the data transfer rate in hydroacoustic channel.

Used for protection of the signals due to pulse and tonal hindrances in majority of modems is OFDM (division of the signal spectrum into sub-bands by the method of short time DFT). Insufficiencies of this method include a rigid connection between width of the frequency channel and buffer duration for its calculation. Therefore, there was proposed another algorithm: division of the spectra into sub-bands with decimation made through polyphase filter, resulting in presence of adaptive equalizer and decoder in each subband. The modem structural scheme is shown in Fig. 6.

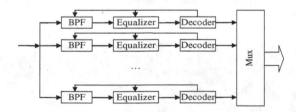

Fig. 6. Structural scheme of hydro-acoustical modem based on poly-phase filters

Use of polyphase modems and equalizer enables to dynamically select number and width of the subbands, as well as the duration of one symbol, resulting in increase of data transfer rate at high SNR, retaining protection against pulse, tonal hindrances and Doppler shift as well as enhancing reliability of receiving data at low SNR through increase of the sub-band width and simplification of the modulation [8]. Obtained through modeling of data transfer was a graph of bit error rate (BER) versus SNR, shown in Fig. 7.

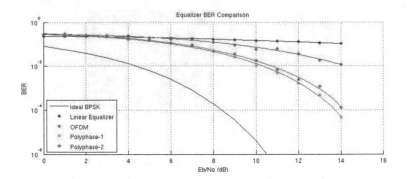

Fig. 7. Modem simulation results in terms of pulse and tone noise

With use of the obtained theoretical results, there was developed a prototype of a hydroacoustic modem based on polyphase filters, which will be tested on an ROV in open water in the summer of 2016.

The stage of the development of electronic equipment of the vehicle included selection of on-board computer, motor control system, power system and a set of gauges.

The function of on-board computer is performed by an open software board Beagle bone Black [19], operating under Linux. The main merits of the board are: small size, low cost and extensive in-built periphery which is of interest for an educational non-commercial project. To connect the sensors and control modules, as well as to protect the board from unfavorable interferences and impulse impact, there has been developed a special extension board featuring the IMU module [19].

The concluding stage of the work included trials, comprising three sub-stages: shore trials, basin trials and open water trials. On shore trials were indispensable from general verification of the functioning of all ROV systems, towing tank trials of the ROV (Fig. 8) permitted to assess dynamics of the vehicle and acceptable operation of the whole system as well as to correlate the results of the trials with the mathematical model developed previously.

Fig. 8. Testing of the experimental ROV model by students of SMTU

Tests conducted in the towing tank of SMTU showed that the experimentally measured dynamic characteristics of the ROV correlate fairly well with those modeled earlier within the Gazebo system.

Intermediate results of the research & education project were presented by the SMTU Ph D student M. Chemodanov and SMTU students' team during the international MATE ROV Competition - 2013 [20].

Open water trials of the ROV are scheduled in the summer of 2016.

4 Conclusion

The SMTU MRER initiative is, on the one hand, orientated toward improvement of the quality of educational process of training of future engineers and providing correlation of the professional competencies required by industrial enterprises with academic competencies acquired by students at the university and, on another hand, - toward the development of innovative technical systems in the field of marine robotics.

The general objectives of the initiative «Marine Robotics: Education through Research» are:

– creation of a «park» of marine robotic platforms of different types to support an innovative research & education process;
– furthering of modular robotic platforms of «open»class«Open-AUV», «Open-ROV», «Open-ASV», which can be upgraded with modules of different useful load for solution of urgent practical problems;
– forming of open research & education resource base on the models of individual and cooperative use of marine robotic platforms;
– setting up of an inter-university Marine Robotics Educational Network for realization of joint research & education projects (including those in the format of double diplomas, structured doctoral programs, research and educational traineeships, etc.)

References

1. Enhanced three-level competency-based curricula in applied marine science. 517271-TEMPUS-1-2011 M-DE-TEMPUS-JPCR (2011)
2. Innovative teaching and learning strategies in open modeling and simulation environment for student-centered engineering education. «Erasmus +» program. 573751-EPP-1-2016-1-DE-EPPKA2-CBHE-JP (2016)
3. Nosov, E.P., Ryzhov, V.A.: Design of the fin propulsor for a submersible propelled by human power. In: Morintekh 2001: Trudi po morskim intellektualnym tekhonologiyam, Saint-Petersburg, 2001, pp. 23–25 (2001). (in Russian)
4. Rozhdestvensky, K.V., Ryzhov, V.A.: Aerohydrodynamics of flapping wing propulsors. Prog. Aerosp. Sci. **39**, 585–633 (2003)
5. Kozhemyakin, I.V., Putintsev, I.A., Semenov, N.N., Chemodanov, M.N.: Development of underwater robotic system, using open-source simulation model extended by hydroacoustic interaction. Izvestiya SFedU. Eng. Sci. **1**(174), 88–102 (2016)
6. Kozhemyakin, I.V., Rozhdestvensky, K.V., Ryzhov, V.A.: The questions of the new gliders generation hydrodynamic design. Materialy devyatoi nauchno-prakticheskoi konferencii «Perspektivnye sistemy i zadachi upravlenya». Taganrog, SFedU, pp. 121–130 (2014). (In Russian)

7. Kozhemyakin, I.V., Potekhin, Yu.P., Rozhdestvensky, K.V., Ryzhov, V.A.: The wave gliders as the marine global information and measurement system elements. In: Materials of 9th R&D Conference «Perspektivnye sistemy i zadachi upravlenya». Taganrog, SFedU, pp. 101–112 (2015). (In Russian)
8. Semenov, N.N.: Choosing probing signal type for active sonar data using the theory of communication channels. Informacionno-upravlyaushie sistemy **1**(37), 47–51 (2009). (In Russian)
9. Intellectual planning of trajectories of moving objects in the media with obstacles. Edited by V.H., Pshihopov. FIZMATGIZLIT, Moscow (2014) (In Russian)
10. Schetz, J.A., Fuhs, A.E.: Fundamentals of Fluid Mechanics. Wiley, New York (1999)
11. Bluerobotics thrusters. https://www.bluerobotics.com/store/thrusters/t200-thruster/
12. Robotic operation system. official site. http://www.ros.org/
13. SMTU ROV software source. https://github.com/MChemodanov/ROV
14. EvoLogics GmbH: official site. https://evologics.de/en/products/index.html
15. Aquatec Ltd.: official site, http://www.aquatecgroup.com/
16. Sonardyne Inc.: official site. http://www.sonardyne.com/products.html
17. LinkQuast Inc.: official site. http://www.link-quest.com/html/intro1.htm
18. Krants V.Z., Sechin V.V.: On increase of transfer speed of the communication system with complex signals in conditions of multi-ray propagation, Materials of the conference «Hydroacoustic communication and hydro-acoustic means ...», Voronezh, Shtil, pp. 216–221 (2007). (In Russian)
19. Beaglebone black, official site. https://beagleboard.org/black
20. MATE ROV competition. http://www.marinetech.org/rov-competition-2/

Human-Machine Speech-Based Interfaces with Augmented Reality and Interactive Systems for Controlling Mobile Cranes

Maciej Majewski[✉] and Wojciech Kacalak

Faculty of Mechanical Engineering, Koszalin University of Technology,
Raclawicka 15-17, 75-620 Koszalin, Poland
{maciej.majewski,wojciech.kacalak}@tu.koszalin.pl

Abstract. In this paper, an overview of human-machine interactive communication for controlling lifting devices is presented, covering also the integration with vision and sensorial systems. Following a general concept, and motivation towards intelligent human-machine communication through artificial neural networks, selected methods are proposed, which provide further directions both of recent as well as of future research on human-machine interaction. The aim of the experimental research is to design a prototype of an innovative interaction system, equipped with a speech interface in a natural language, augmented reality and interactive manipulators with force feedback.

Keywords: Speech communication · Intelligent interface · Interactive system · Natural language processing · Neural networks · Intelligent control · Human-machine interaction · Artificial intelligence

1 The Design of an Innovative Human-Machine Interface

Innovative ARSC (Augmented Reality and Smart Control) systems designed for processes of precise positioning of objects and cargo can be equipped with intelligent systems of speech interaction between lifting devices and their human operators. Artificial intelligence-based technologies find their application in modern systems for controlling and supervising machines using vision systems - machine vision [1], augmented reality [2], speech-based interactive communication [3–7] as well as interactive controllers [8] providing force feedback.

The design and implementation of human-machine interactive systems is an important field of applied research. Recent advances in development of prototypes of speech-based interfaces are described in articles in [4,9–11]. The proposed system is presented in abbreviated form in Fig. 1. The concept specifies integration of a system for natural-language communication with visual and sensorial systems. The presented research involves the development of a system for controlling a loader crane, equipped with vision and sensorial systems, interactive manipulators with force feedback, as well as a system for bi-directional

© Springer International Publishing Switzerland 2016
A. Ronzhin et al. (Eds.): ICR 2016, LNAI 9812, pp. 89–98, 2016.
DOI: 10.1007/978-3-319-43955-6_12

communication through speech and natural language between the operator and the controlled lifting device.

The fundamental concept of the interaction systems between loader cranes and human operators assumes that they are equipped with the following subsystems: augmented reality vision, speech communication, natural language processing, command effect analysis, command safety assessment, command execution, supervision and diagnostics, decision-making and learning, interactive manipulation with force feedback. The novelty of the system also consists of inclusion of several adaptive layers in the spoken natural language command interface for human biometric identification, speech recognition, word recognition, sentence syntax and segment analysis, command analysis and recognition, command effect analysis and safety assessment, process supervision and human reaction assessment. The proposed interactive system (Fig. 2) contains many specialized modules and it is divided into subsystems.

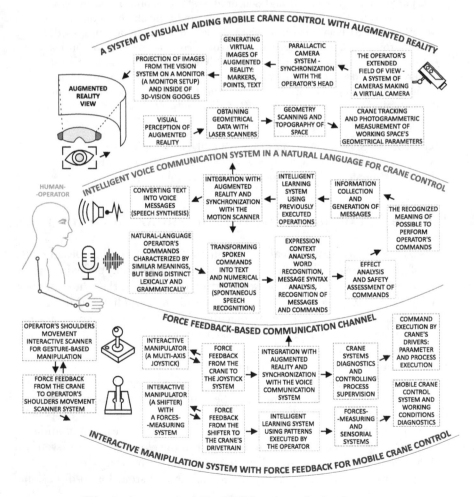

Fig. 1. A concept of the ARSC systems for loader cranes.

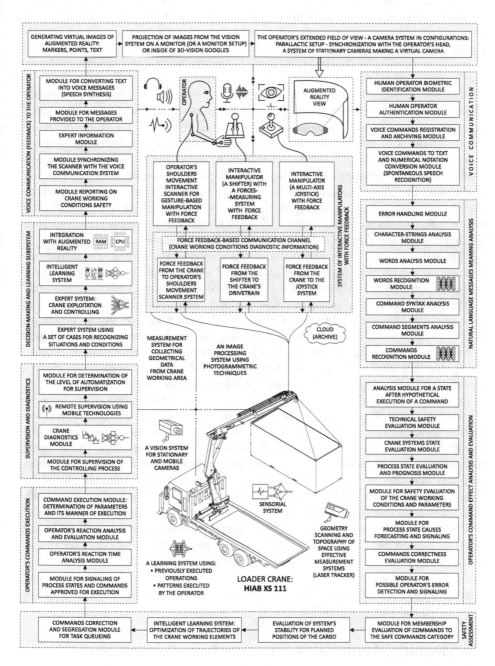

Fig. 2. Structure of the system for speech interaction between the loader crane (Hiab XS 111) and the operator equipped with vision systems and interactive manipulators.

The subsystem of visually aiding loader crane control with augmented reality generates virtual images of augmented reality (including markers, points), and also projects images from the vision system on a monitor setup or inside of 3D-vision goggles. The operator's extended field of view contains a camera system in configurations: parallactic setup - synchronization with the operator's head, and a system of stationary cameras making a virtual camera.

The subsystem for speech communication is used to perform the following tasks: processing the operator's spoken commands, operator biometric identification, converting voice commands to text and numerical notation, handling errors, analysis of words, recognition of words, analysis of commands' syntax, analysis of commands' segments, recognition of commands, meaning analysis of natural-language messages, as well as converting text into voice messages (speech synthesis). The voice communication subsystem also provides voice feedback to the operator including reporting on the crane's working conditions' safety and expert information for exploitation and controlling. It is also communicated with a subsystem of interactive manipulators with force feedback.

The subsystem of effect analysis and evaluation of the operator's commands is designed for the following tasks: analysis of a state after hypothetical execution of a command, evaluation of technical safety, evaluation of crane systems' and process's states, evaluation of crane working conditions' safety, forecasting of process states' causes, evaluation of commands' correctness, as well as detection of the operator's errors. The commands' safety assessment subsystem is assigned to evaluate the command correctness. The subsystem of the command execution is capable of signaling of process states. The execution of commands involves determination of process's parameters and its manner of execution for the configuration of the crane. The subsystem for supervision and diagnostics implements crane diagnostics, supervision of the controlling process, remote supervision with mobile technologies. It also includes the tasks related to measurements of the crane's working space and collection of geometrical data using photogrammetric techniques. The decision-making and learning subsystem is composed of expert systems, and the intelligent learning kernel integrated with augmented reality. The system is also linked with the interactive manipulators providing force feedback, which include the operator's shoulders' movement interactive scanner for gesture-based manipulation, a shifter with the forces-measuring system, and a multi-axis joystick. It is a connection to a force feedback-based communication channel (crane's working conditions diagnostic information) containing force feedback from the crane to the operator's shoulders' movement scanner system, force feedback from the shifter to the crane's drivetrain, as well as force feedback from the crane to the joystick system.

2   Meaning Analysis of Commands and Messages

The concept of the ARSC system includes a subsystem of recognition of speech commands in a natural language using patterns and antipatterns of commands, which is presented in Fig. 3. In the subsystem, the speech signal is converted to text and numerical values by the continuous speech recognition module. After a successful utterance recognition, a text command in a natural language is further processed. Individual words treated as isolated components of the text are subsequently processed with the modules for lexical analysis, tokenization and

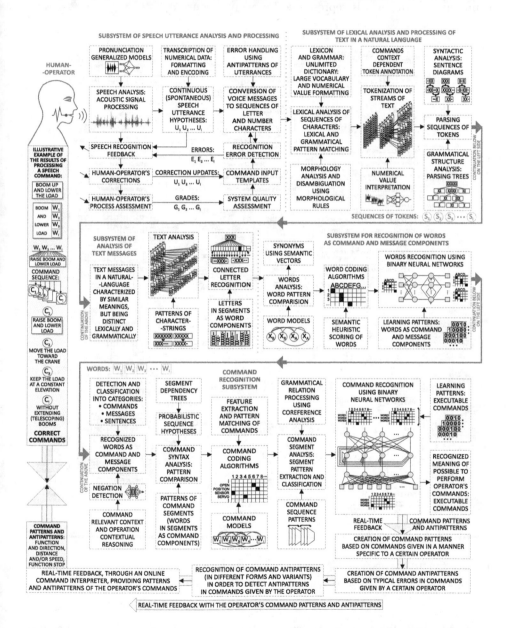

Fig. 3. A concept of a system of recognition of speech commands in a natural language using patterns and antipatterns of commands.

parsing. After the text analysis, the letters grouped in segments are processed by the word analysis module. In the next stage, the analyzed word segments are inputs of the neural network for recognizing words. The network uses a training file containing also words and is trained to recognize words as command components, with words represented by output neurons.

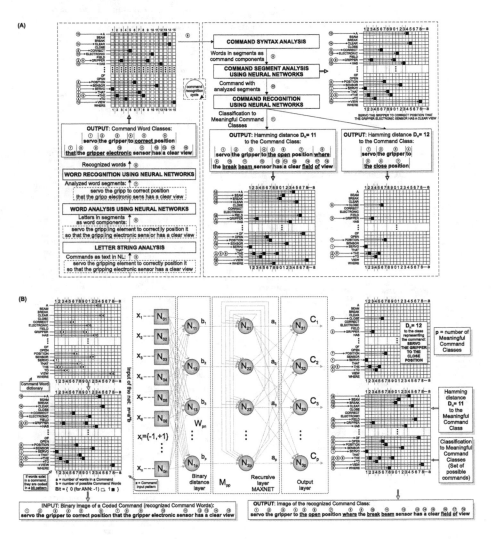

Fig. 4. (A) Block diagram of a meaning analysis cycle of an exemplary command, (B) Illustrative example of recognition of commands using binary neural networks.

In the meaning analysis process of text commands (Fig. 4A) in a natural language, the meaning analysis of words as command or message components is performed. The recognized words are transferred to the command syntax analysis module which uses command segment patterns. It analyses commands and identifies them as segments with regards to meaning, and also codes commands as vectors. They are sent to the command segment analysis module using encoded command segment patterns. The commands become inputs of the command recognition module. The module uses a 3-layer Hamming network to classify the command and find its meaning (Fig. 4B). The neural network of this module uses a training file with possible meaningful commands.

The proposed method for meaning analysis of words, commands and messages uses binary neural networks (Fig. 4) for natural language understanding. The motivation behind using this type of neural networks for meaning

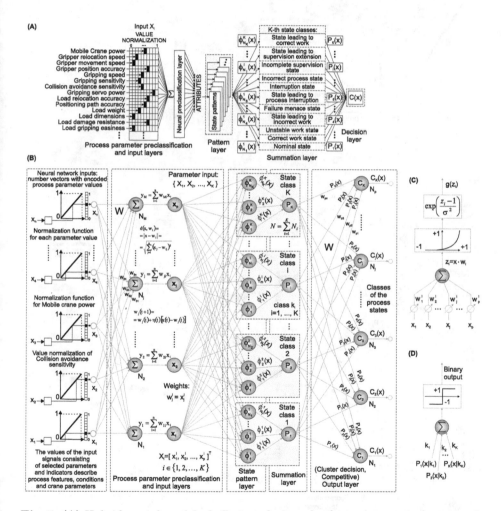

Fig. 5. (A) Hybrid neural model of effect analysis and safety assessment of commands in a cargo manipulation process, (B) The architecture of the hybrid neural network used, (C) Neuron of the pattern layer, (D) Neuron of the output layer.

analysis [12] is that they offer an advantage of simple binarization of words, commands and sentences, as well as very fast training and run-time response. The cycle of exemplary command meaning analysis is presented in Fig. 4A. The proposed concept of processing of words and messages enables a variety of analyses of the spoken commands in a natural language.

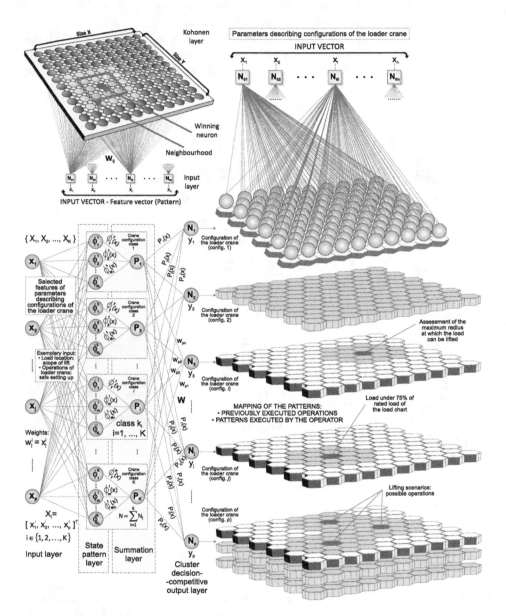

Fig. 6. Proposed learning systems using previously executed operations and patterns executed by the operator.

3   Effect Analysis and Safety Assessment of Commands

In the innovative speech interface, the problem of effect analysis and safety assessment of commands can be solved with hybrid neural networks. The proposed method (Fig. 5A) uses developed hybrid multilayer neural networks consisting of a modified probabilistic network combined with a single layer classifier. The probabilistic network is interesting, because it is possible to implement and develop numerous enhancements, extensions, and generalizations of the original model [13]. The effect analysis and safety assessment of commands is based on information on features, conditions and parameters of the cargo positioning process. The effect analysis and safety assessment is performed by the developed hybrid network that works as a classifier of the cargo manipulation process state. The architecture of the proposed hybrid network is shown in Fig. 5B–D.

The proposed innovative speech interface is equipped with learning) systems using previously executed operations and patterns executed by the operator. The developed learning systems are based on proposed hybrid neural networks (Fig. 6) consisting of self-organizing feature maps (Kohonen networks [14] combined with a probabilistic classifier. The inputs of the hybrid networks contain selected features of the parameters describing configurations of the loader crane. The outputs represent individual configurations of the crane which provide self-organizing feature maps of the previously executed operations and patterns executed by the operator.

4   Conclusions and Perspectives

The designed interaction system is equipped with the most modern artificial intelligence-based technologies: voice communication, vision systems, augmented reality and interactive manipulators with force feedback. Modern control and supervision systems allow to efficiently and securely transfer, and precisely place materials, products and fragile cargo. The proposed design of the innovative AR speech interface for controlling lifting devices has been based on hybrid neural network architectures. The design can be considered as an attempt to create a new standard of the intelligent system for execution, control, supervision and optimization of effective and flexible cargo manipulation processes using communication by speech and natural language.

Acknowledgements. This project is financed by the National Centre for Research and Development, Poland (NCBiR), under the Applied Research Programme - Grant agreement No. PBS3/A6/28/2015.

References

1. Pajor, M., Grudziski, M.: Intelligent machine tool - vision based 3D scanning system for positioning of the workpiece. Solid State Phenom. **220–221**, 497–503 (2015)

2. Pajor, M., Miadlicki, K.: Real-time gesture control of a CNC machine tool with the use Microsoft Kinect Sensor. Int. J. Sci. Eng. Res. **6**(9), 538–543 (2015)
3. Kacalak, W., Majewski, M., Budniak, Z.: Intelligent automated design of machine components using antipatterns. In: Jackowski, K., Burduk, R., Walkowiak, K., Wozniak, M., Yin, H. (eds.) IDEAL 2015. LNCS, vol. 9375, pp. 248–255. Springer, Heidelberg (2015). doi:10.1007/978-3-319-24834-9_30
4. Kacalak, W., Majewski, M., Budniak, Z.: Interactive systems for designing machine elements and assemblies. Manag. Prod. Eng. Rev. **6**(3), 21–34 (2015). De Gruyter Open
5. Kacalak, W., Majewski, M.: New intelligent interactive automated systems for design of machine elements and assemblies. In: Huang, T., Zeng, Z., Li, C., Leung, C.S. (eds.) ICONIP 2012, Part IV. LNCS, vol. 7666, pp. 115–122. Springer, Heidelberg (2012)
6. Kacalak, W., Majewski, M., Zurada, J.M.: Intelligent e-Learning Systems for evaluation of user's knowledge and skills with efficient information processing. In: Rutkowski, L., Scherer, R., Tadeusiewicz, R., Zadeh, L.A., Zurada, J.M. (eds.) ICAISC 2010, Part II. LNCS, vol. 6114, pp. 508–515. Springer, Heidelberg (2010)
7. Kacalak, W., Stuart, K.D., Majewski, M.: Intelligent natural language processing. In: Jiao, L., Wang, L., Gao, X., Liu, J., Wu, F. (eds.) ICNC 2006. LNCS, vol. 4221, pp. 584–587. Springer, Heidelberg (2006)
8. Pajor, M., Miadlicki, K.: Overview of user interfaces used in load lifting devices. Int. J. Sci. Eng. Res. **6**(9), 1215–1220 (2015)
9. Majewski, M., Kacalak, W.: Intelligent speech interaction of devices and human operators. In: Silhavy, R., et al. (eds.) CSOC 2016. AISC, vol. 465, pp. 471–482. Springer, Cham (2016)
10. Kumar, A., Metze, F., Kam, M.: Enabling the rapid development and adoption of speech-user interfaces. Computer **47**(1), 40–47 (2014). IEEE
11. Ortiz, C.L.: The road to natural conversational speech interfaces. IEEE Internet Comput. **18**(2), 74–78 (2014). IEEE
12. Majewski, M., Zurada, J.M.: Sentence Recognition using artificial neural networks. Knowl. Based Syst. **21**(7), 629–635 (2008). Elsevier
13. Specht, D.F.: Probabilistic neural networks. Neural Netw. **3**(1), 109–118 (1990). Elsevier
14. Kohonen, T.: Self-Organization and Associative Memory. Springer, Heidelberg (1984)

Human-Robot Interaction Using Brain-Computer Interface Based on EEG Signal Decoding

Lev Stankevich and Konstantin Sonkin[✉]

Peter the Great Saint Petersburg Polytechnic University, 29, Polytechnicheskaya,
St. Petersburg 195251, Russia
Stankevich_lev@inbox.ru, sonkinkonst@mail.ru

Abstract. This study describes a new approach to a problem of the human-robot interaction for remote control of robot behavior. Finding a solution to this problem is important for providing control of robots and unmanned vehicles. At the interaction a human operator can form commands for robot control. It is proposed to use a noninvasive brain-computer interface based on the decoding of signals of brain activity during motor imagery to generate the supervisor commands for robot control. The principles of the interaction of human as an operator and robot as an executor are considered. Using the brain-computer interface the operator can change robot behavior without any special movements and modules embedded into robot's program. The study aimed to development of the human-robot interaction system for non-direct control of the robot behavior based on the brain-computer interface for classification of EEG patterns of imaginary movements of one hand fingers in real-time. Example of such human-robot interaction realization for Nao robot with neurofeedback is considered.

Keywords: Brain-computer interface · Classifier committee · Imaginary finger movements · Human-robot interaction

1 Introduction

The human future is directly connected to the development of means of human-machine communications. At the first stage of the development people communicated with computers using traditional means of the communications, which included keyboards, monitors, printers, devices of graphical input and output realizing simple interfaces. During the artificial intelligence development, speech and visual communication means have been created for intellectual interfaces. In the last 20 years rapid development of neurophysiology, artificial intelligence and methods and devices of registration of bioelectrical signals have provided the possibility of elaboration of new means of human-machine communications based on direct perception of nervous system signals.

Nowadays among the most popular machines controlled by computers, robots are widely used as assistants of people. These can be mobile robots participating in rescue and military operations or unmanned vehicles. Currently strong interest arises to a problem of autonomous control of robots with service functions. At the present time interaction of disabled human and humanoid robot assistant is the most urgent. The

© Springer International Publishing Switzerland 2016
A. Ronzhin et al. (Eds.): ICR 2016, LNAI 9812, pp. 99–106, 2016.
DOI: 10.1007/978-3-319-43955-6_13

humanoid robot is the new kind of mobile robots with human-like form and behavior. That's why development of the user-friendly human-robot interface is crucial for the effective application of such robots for rehabilitation and care of disabled people.

The perspective way of realization of the user-friendly human-robot interaction is the use of so-called brain-computer interfaces (BCI). BCI is a modern neural technology, capable to provide human communications with external electronic and electromechanical devices on the basis of registration and decoding of brain activity signals. BCI allows a human to cooperate with environment by means of transformation of brain activity signals into control commands for external devices or computer programs [1]. Such BCI should work in real time and therefore to form control signals with the minimum delay defined by the speed of external device's feedback. BCI can be used for rehabilitation of patients after stroke and in other cases. For rehabilitation purposes BCI based on recognition of imaginary movements is one of the most suitable [2].

Electroencephalography (EEG) is often used as the mean of registration of the bioelectric activity of brain corresponding to motor commands. Modern researches show that EEG has considerable potential for implementation in BCI as noninvasive and rather low-cost technique. However, the factor restricting practical application of BCI based on EEG signals is the complexity of reliable and stable decoding of brain signals. Another factor is the difficulty of classification of EEG patterns of imaginary movements in a real time.

At the present time noninvasive BCIs based on EEG are realized for decoding of imaginary and real movements of large parts of body, for example, arms and feet. At the same time such BCIs are not effective for decoding of fine movements, for example, fingers of one hand. This problem is difficult due to the anatomic affinity of the brain structures involved in the implementation of imaginary movements, and insignificant differences in EEG signals of imaginary movements of fine body parts. The solution of this problem demands big computing resources for data analysis. It is necessary to note that for real time system there are additional requirements for multithreading and time delays.

There are several scientific groups and the organizations in Russia that conduct studies in the field of BCI development, many of which are known at the international level [3, 4]. Thus, there are attempts for creating human-robot interface using BCI based on EEG signals.

The most of studies devoted to control of robots at individual and group levels. At the individual level there is interaction of human operator and a single robot as a separate act in which operator forms a message with instructions (commands) for robot and checks their executions. At the group level of interaction the special interest presents a problem of interaction of operator and the group of robots, performing teamwork, which requires carrying out common intentions of all team members to achieve the common goal.

The objective of the study is the development of the human-robot interaction system for non-direct control of robot behavior by means of the BCI based on online EEG pattern classification of imaginary movements of fingers with minimal time delay.

In the following sections principles of the human-robot interaction using BCI for robot control system organization are discussed. In the third section the real time BCI is described. In the fourth section experiments related to BCI robot control are described. The work is based on authors' experience in design of noninvasive BCI and intelligent control systems of robots.

2   Principles of Human-Robot Interaction Using BCI

In robot control systems operators must have possibility to input instruction to change behavior of robots. In this work it is proposed to do it through noninvasive BCI decoding brain activity signals and forming supervisor commands for changing robot behavior. For practical realization of such a system neurofeedback can be used, which allows the brain to adapt and to enhance EEG signals of imaginary commands [3]. The delay of the system response should be small enough in order to provide coupling of brain signal and performed command in user consciousness.

For implementation of the human-robot interaction systems using noninvasive BCI it is necessary to provide the effective interacting of systems of registration, preprocessing, classification of EEG signals, and control (based on the decoded motor command) in real-time.

Basic constructing principles of such systems of human-robot interaction are related to realization of BCI in real-time and could be formulated as follows:

- achievement of high accuracy of EEG pattern classification, which is acceptable for effective usage of BCI in robot control systems;
- providing enough degrees of freedom for BCI, which is determined by the amount of the decoded imaginary commands; for example, the most of realized BCI classify 3–4 mental states [5–8];
- optimization of time and computing resources required for feature extraction, using single trial approach [9, 10].

3   Structure and Components of Real Time BCI

Noninvasive BCIs based on EEG include modules of registration and preprocessing of EEG signals for features extraction in spectral or time domains, EEG pattern classification of imaginary commands and command recognition (Fig. 1).

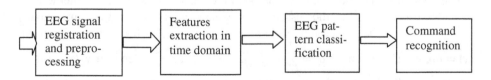

Fig. 1. Structure of noninvasive BCI

To study the possibility of implementation of the human-robot interaction system based on noninvasive BCI, the variant of BCI earlier developed by authors was applied for classification of imaginary movements of fingers of one hand [11, 12]. Adaptation of this variant of BCI to robot control was made by the use of specially developed program and methodic. According to the methodic we offered subjects to imagine 4 types of imaginary movements of fingers, which corresponded to commands for robot: "Forward", "Stop", "To the right", "To the left". Each of the commands was matched with the movement of the certain finger of the right hand - thumb, index, middle, and little fingers accordingly. On the first step subject should push the button of the computer mouse by the finger appointed for control of the required robot behavior, in rhythm given by sounds. When the sound ceased, subject had to continue to *imagine* movements by the same finger (without real movements). The instruction is oriented on imagination of movements with kinetic static feelings [13]. In one block of trials real and imaginary movements repeated several times. As a result the subject should execute not less than hundred real and hundred imaginary movements in given rhythm in one block of the experiment.

Registration of EEG was carried out by means of 32 channel digital EEG encephalograph by "Mitsar" and the software package WinEEG nineteen silver-chloride scalp electrodes that were located according to the 10–20 international system. EEG signals were registered in the frequency band 0.53 Hz–30 Hz. The electrode impedance was kept at less than 5 kΩ, the quantization frequency was 500 Hz. EEG artifacts such as movements of eyes were excluded from the analysis. The time area of the analysis was 600 ms from the beginning of trials. In this window the last stages of preparation to imaginary movement and the imaginary movement itself were presumably located.

Analysis of EEG patterns was made in time domain. To maintain the necessary speed of the system feature extraction was carried out from each separate trial without accumulation (single-trial approach). The algorithm of the joint analysis of two feature spaces was used: the square under the curve and the length of the curve of segments of EEG signal. The given features were computed in sliding window of analysis.

The developed committee of qualifiers of EEG patterns is based on artificial neural networks (ANN) and support vector machines (SVM). These methods are effective means of classification of imaginary movements [14].

ANNs are based on principles of the nonlinear, distributed, parallel and local data processing and adaptation. In this work ANNs are realized in the form of multilayered perceptron, consisting of three layers: two hidden and one output. As activation functions in the hidden layers sigmoid functions were used. Linear function was used in the output layer.

The method of SVM offered by Cortes and Vapnik [15] is related to methods of linear classification. The method essence consists in separation of sample into classes by means of the optimum dividing hyperplane which has equation: $f(x) = (\omega, \phi(x)) + b$, where $\omega = \sum_{i=1}^{N} \lambda_i y_i(x_i)$, coefficients λ_i depend on y_i (vectors of class membership labels) and from values of scalar products. Thus, it is required to know the values of scalar products, which are determined by the kernel function: $k(x, y) = (\varphi(x), \varphi(y))$.

Based on the results of the studies [12, 15, 16] in our work the SVM classifier based on a Gaussian radial basis function: $K(x_i, y_j) = \exp\left(-\gamma \|x_i - x_j\|^2\right)$, for $\gamma > 0$ was implemented using the MATLAB LIBSVM package [17].

The two level classifier committee has been developed. It has the first level consisting of two ANNs (ANN (S) and ANN (L)) and two SVM based qualifiers (SVM (S) and SVM (L)), and the second level the second-level generalizing neural network (ANN (C)) for merging the results of the first level qualifiers (Fig. 2).

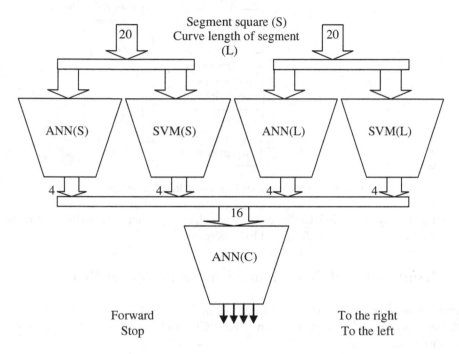

Fig. 2. Structure of the two level classifier committee

Classifiers of the first level analyze separate feature vectors (the length of the curve L or the square under the curve S) obtained for 20 segments of EEG signal (20 features on L and 20 features on S). The results of the first-level classifiers form the feature vector for the second level generalizing ANN (16 components), which makes the final decision on the assigning of the trial to one of the classes.

Work of BCI in the real-time regime assumes continuous reading of EEG data. To maintain the continuum of an input data recording and its parallel analysis, the method of multi-threaded programming has been used: data flow 1 records EEG in the buffer and data flow 2 debarks data from the buffer and process it (Fig. 3).

The system presented on Fig. 3 works as follows: as soon as the trial N has ended and the following trial N + 1 starts, data flow 2 debarks the data of the trial N, pre-process it, extracts features and classifies the obtained EEG pattern. Thus, simultaneous

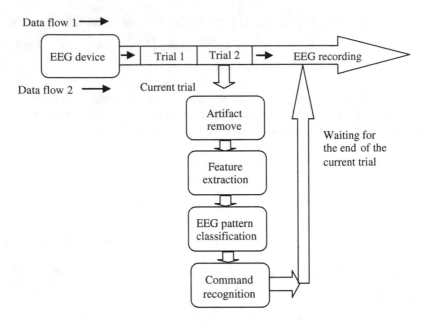

Fig. 3. Scheme of the real-time classification system

record and analysis of EEG data is realized, which is important for classification of all trials without exceptions and for speed BCI increase.

4   Results of Robot Control Based on the Real-Time BCI

Anthropomorphic robot NAO with 23 degrees of freedom (Fig. 4) was used for robot control experiments by means of noninvasive BCI with the system of classification of EEG patterns.

Fig. 4. Mobile robot controlled by noninvasive BCI

This robot was controlled by means of BCI, which formed the high level (supervisor) commands instead of the direct control of drives of degrees of freedom. Four commands were used: "Forward", "Stop", "To the right", "On the left". The experiment showed possibility to train the classifiers of BCI to recognize these commands and to use them for the robot control.

The conducted research of BCI prototype has shown that its characteristics meet the requirements of real-time control systems. Time delays required to receive the answer from the classifier are in limits of 600 ms that is sufficiently small to use in BCI with neurobiofeedback. According to the self reports of subjects their imaginary commands were clearly linked to the obtained answers of the control system.

The probability of the successful classification of imaginary commands in BCI is highly individual. As shown a certain part of subjects (32.6 %) cannot work with BCI based on EEG signals. In this study the average decoding accuracy of 4-class classification by means of the committee of heterogeneous classifiers was 60 ± 10 %, maximum - 77 ± 5 % (guessing level for the 4-class problem was 25 %).

5 Conclusions

The study highlighted several problems of elaboration of the effective system of human-robot interaction based on noninvasive real-time BCI such as low accuracy and speed of classification of EEG patterns of imaginary movements, small number of degrees of freedom and demand in optimization of computing resources.

Noninvasive BCI, which met the requirements of practical use in real time, was developed. The maximum delay of the system did not exceed the duration of one trial that allowed using neurobiofeedback. Multithreading realized in the system allowed to record and analyze EEG signal simultaneously without data loss and to carry out reliable on-line classification.

Efficiency of the developed BCI could be additionally raised by the training of subjects with the use of neurofeedback, enhancement of mathematical methods of features extraction and by the individual adjustment of the classification system for subjects. In the future studies it's supposed to develop the human-robot interaction modules for both individual and group robot control in real time.

References

1. Wolpaw, J.R., Wolpaw, E.W.: Brain-Computer Interfaces: Principles and Practice. Oxford University Press, Oxford (2012)
2. Daly, I., Billinger, M., Laparra-Hernández, J., Aloise, F., García, M.L., Faller, J., Scherer, R., Müller-Putz, G.: On the control of brain-computer interfaces by users with cerebral palsy. Clin. Neurophysiol. **124**, 1787–1797 (2013)
3. Frolov, A.A., Roshin, V.U.: Brain computer interface. Reality and perspectives. In: Scientific Conference on Neuroinformatic MIFI 2008. Lections on Neuroinformatics (2008). http://neurolectures.narod.ru/2008/Frolov-2008.pdf (in Russian)

4. Kaplan, A.Y., Kochetkov, A.G., Shishkin, S.L., et al.: Experimental-theoretic bases and practical realizations of technology "Brain computer interface". Sibir Med. Bull. **12**(2), 21–29 (2013). (in Russian)

5. Bai, O., Lin, P., Vorbach, S., Floeter, M.K., Hattori, N., Hallett, M.: A high performance sensorimotor beta rhythm-based brain-computer interface associated with human natural motor behavior. J. Neural Eng. **5**(1), 24 (2008)

6. Hsu, W.: Embedded grey relation theory in hopfield neural network application to motor imagery eeg recognition. Clin. EEG Neurosci. **44**(4), 257–264 (2013)

7. Huang, D., Lin, P., Fei, D.Y., Chen, X., Bai, O.: EEG-based online two-dimensional cursor control. In: Conference of the IEEE Engineering in Medicine and Biology Society, pp. 4547–4550 (2009)

8. Leeb, R., Scherer, R., Keinrath, C., Guger, C., Pfurtscheller, G.: Exploring virtual environments with an EEG-based BCI through motor imagery. Biomed. Technik. **52**, 86–91 (2005)

9. Asensio-Cubero, J., Gan, J.Q., Palaniappan, R.: Multiresolution analysis over graphs for a motor imagery based online BCI game. Comput. Biol. Med. **68**(1), 21–26 (2016)

10. Billinger, M., Brunner, C., Müller-Putz, G.R.: SCoT: a Python toolbox for EEG source connectivity. Front. Neuroinformatics **8**, 22 (2014)

11. Sonkin, K.M., Stankevich, L.A., Khomenko, J.G., Nagornova, Z.V., Shemyakina, N.V.: Development of electroencephalographic pattern classifiers for real and imaginary thumb and index finger movements of one hand. Artif. Intell. Med. **63**(2), 107–117 (2015)

12. Stankevich, L.A., Sonkin, K.M., Shemyakina, N.V., Nagornova, Z.V., Khomenko, J.G., Perts, D.S., Koval, A.V.: Pattern decoding of rhythmic individual finger imaginary movements of one hand. Hum. Phisiology **42**(1), 32–42 (2016)

13. Neuper, C., Scherer, R., Reiner, M., Pfurtscheller, G.: Imagery of motor actions: differential effects of kinesthetic and visual-motor mode of imagery in single-trial EEG. Cogn. Brain. Res. **25**, 668–677 (2005)

14. Lotte, F., Congedo, M., Lecuyer, A., et al.: Review of classification algorithms for EEG-based brain-computer interfaces. J. Neural Eng. **4**(2), 1 (2007)

15. Cortes, C., Vapnik, V.N.: Support-vector networks. Mach. Learn. **20**(3), 273 (1995)

16. Shawe-Taylor, J., Cristianini, N.: Kernel Methods for Pattern Analysis. Cambridge University Press, New York (2004). http://www.kernel-methods.net

17. Chang, C.C., Lin, C.J.: LIBSVM: a library for support vector machines. ACM Trans. Intell. Syst. Technol. **2**(27), 1–27 (2011). http://www.csie.ntu.edu.tw/~cjlin/libsvm

Interactive Collaborative Robotics and Natural Language Interface Based on Multi-agent Recursive Cognitive Architectures

Murat Anchokov, Vladimir Denisenko, Zalimkhan Nagoev,
Zaurbek Sundukov[✉], and Boris Tazhev

Institute of Computer Science and Problems of Regional Management of Kabardino-Balkarian
Scientific Center of Russian Academy of Sciences,
I.Armand Street, 37a, Nalchik, Russia
iipru@rambler.ru, azraiths@gmail.com

Abstract. The article represents a review of the world current state in robotics, its fields of implementation and application. It provides a description of a complex collaborative robotic system for monitoring and reconnaissance of leaks and spills of flammable, explosive and toxic substances, for elimination and prevention of accidents aftermaths. The work describes the robotic system software and functional modules. It introduces methods and algorithms of natural language interaction on the basis of multi-agent recursive cognitive architecture.

Keywords: Collaborative robot · Robotic system · Natural language interface · Program control station · Monitoring and reconnaissance

1 Introduction

Practical robotics is no more a category of scientific or engineering interests but of commercial interests. In a report, Worldwide Commercial Robotics Spending Guide, it is forecasted that the compound annual growth rate (CAGR) in robotics and related services will grow 17 % per year, in other words, it will increase from $ 71 billion in 2015 to $ 135.4 in 2019 [1, 2].

In contrast to previous technological revolutions, the transition to a new technological mode, characterized by massive substitution of physical and mental labor for unmanned execution of production and management actions by robots, takes place in the context of formed solvent demand on new equipment.

Human life will change dramatically because of robot implementation in houses, all services, hobby and spare time etc. Speedy (in 3–4 years) implementation of robots is expected in the consumer and service sectors: warehouse and transport logistics, various infrastructure inspection, entertaining business, marketing and trade, delivery services, agriculture, medicine, and construction. The sphere of education and social services, security services, housing and communal services (directly or indirectly) have already been searching for advanced systems that can replace human staff.

© Springer International Publishing Switzerland 2016
A. Ronzhin et al. (Eds.): ICR 2016, LNAI 9812, pp. 107–112, 2016.
DOI: 10.1007/978-3-319-43955-6_14

There are also radical changes in the system of robotics offers. Advances in the development of intellectualization of information processing and decision-making lead to the rapid expansion of the applicability of robots and to a dynamic market growth. Current developments are oriented to the use of "human" tools, industrial and household equipment, ensuring safe transfer in industrial and home environment. According to the International Federation of Robotics, the population of personal robots will reach 100 million units within five years. A number of countries have begun researches on the impact of robotics technological on the legal and ethical standards, the infrastructure of cities and strategic communications.

Collaborative robots is a relatively new trend in robotics that supposes side-by-side man-machine co-operation. Collaborative robots are usually equipped with sensors and machine vision to avoid robot-human collisions. These robots are not supposed to be used in specially shielded areas but in close cooperation with people. The main task of such side-by-side collaboration is to protect people from injury.

Let us take a close look at the state of the art in collaborative robotics.

ABB YUMI is a double-handed robot, its main application field is small parts assembly. This robot is rather compact, has accurate vision, dexterous grippers, sensitive force control feedback, flexible software and built-in safety features [3].

Roberta robot is designed for small and medium-sized enterprises. The robot's main characteristics are: safe, lightweight, 6 degrees of freedom without singularity points, good pay load to structural weight ratio, cheap. It is equipped with cameras that capture abnormal items in the gripper. The machine vision is not just for safety, it is used to guide the manipulator to the right position as well [4].

APAS is a first certificated assistant robot. It is in a protection leather cover that provides direct man-machine cooperation without safety fences [5].

FANUC has designed the world first collaborative robot with a 35 kg payload. There are differences between CR-35iA and the rest of the FANUC family: it is green, it has soft shell, it is not surrounded by defensive shields. That frees up a lot of floor space und cuts costs. It will gently stop if someone comes too close while it is working [6].

F&P Personal Robotics has launched ultra-light-weight PRob1R. It can be used in multi-step assembly tasks because it is fast and safe and due to its weight can be quickly modified to different sizes and types of batches [7].

There are many other collaborative robots that are worth paying attention at, for instance, NEXTAGE, KUKA - IIWA, MABI SPEEDY-10, MRK SYSTEME KR 5 SI, PAVP6, Scara, etc.

2   Problem Formulation

The aim of this work is a description of the software of the prototype of multi-agent robotic system. The robot is designed for monitoring leaks and spills of flammable, explosive and toxic substances, fires, situation analysis, decision-making, synthesis and implementation of collective action plans, prevention and elimination of accidents after-maths, catastrophes and natural disasters on the basis of interaction with the stationary robotic firefighting complexes in conjunction with mobile technical means of reconnaissance and hardware to optimize decision-making and carrying out fire and rescue operations in the complex conditions.

3 Implementation

The functional analysis of the robotic system allows to identify the following functional modules:

- multi-agent system of aggregation of data from different measuring devices and the formation of a unified model of the emergency area on local data registered by a distributed robotic system and stationary sensors;
- intelligent decentralized control expert system of decision-making under conditions of accidental leaks, spills and fires on the basis of a distributed multi-agent knowledge base;
- station software wearable of management and control;
- software modules of the robotic system.

Portable command station includes the operator workstation, equipped with a digital radio link with modules of the robotic system, and a computer with a special software interface, supporting the high-level exchange of data, analyzing and visualizing data on the status and conditions of complex tasks.

Software control center contains the following components:

- video module for teleoperator mode;
- module of graphic realization of multimodal data;
- module that provides interaction with the operator and robot via natural language interface.

Figure 1 represents the video module interface that provides the following functions:

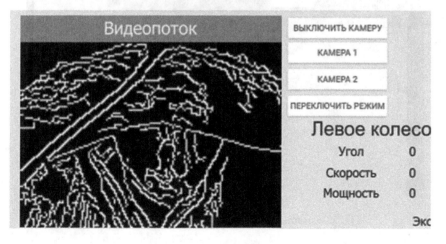

Fig. 1. The video module recognition at work

- receive and display data from the camera;
- provide with correct data;
- change the module operating modes of onboard computer vision.

Figure 2 presents module of graphical visualization, providing the following functions:

- receive information from sensory subsystems;
- visualize received information;
- provide an opportunity to control the robot by means of control commands from the keyboard and gamepad devices.

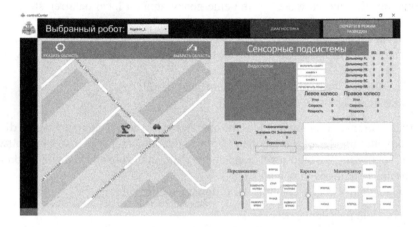

Fig. 2. The main window of the module

Data visualization module updates information in real time on the map, and allows to observe emergency boundaries (Fig. 3).

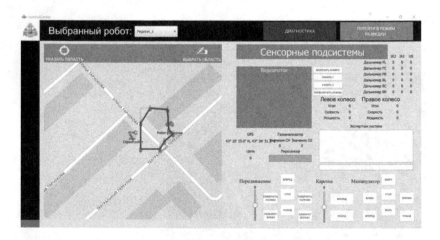

Fig. 3. Construction of contours of the emergency zone

4   Natural Language Module

The module is a software system, its components are on the command station and on-board computer of the autonomous mobile robot. The system provides a bi-directional exchange of natural language statements in a dialog system style and is designed to input tasks (missions) into the intelligent control system of the robot. The dialog mode is used to clarify and confirm the understanding of structure and sequencing of missions, limited by the application area of the autonomous robot.

Natural language interaction is based on multi-agent recursive cognitive architecture (MuRCA) [8–14]. MuRCA allows to understand the limited subset of natural language in the following way: each word has agents for its representation in the intelligence system. A "working life" of an agent depends on the amount of energy it possesses, i.e. the lower it is, the higher are its chances to "die". In order to support its life, from the very beginning of its creation agent is determined to make contracts with other agents that already exist in the system.

These contracts have two main functions: energy-information exchange and representation of linguistic interrelationship between agents. Contracts can be made only between agents that make sense, in other words, there cannot be a contract between "dark" and "go", because there is no such syntactic structure in the natural language; but there can be a contract between "dark" and "spot" or "go" and "we", because there are such constructions in the natural language.

According to this algorithm, we can teach the system words of the limited subset of the natural language that is necessary to the robot. Consider that the system already knows words: "turn" and "left". Let us teach it the phrase "turn left". After we have typed or pronounced the phrase, the agents in the system that are responsible for these separate words activate and send each other offers to make contracts.

After all necessary contracts are made, the system makes an attempt to execute the order. The robot carries out an action or a set of actions and the system asks the user whether this action is right. If it is right, the system remembers the phase and the corresponding action, if it is wrong, the system asks for some specifications and the procedure cycles.

Fig. 4. A complex collaborative robotic system

5 Conclusions

The result of this work is a complex collaborative robotic system that monitors and reconnoiters the emergency zone, prevent and eliminate accidents aftermaths and disasters (Fig. 4), program control station, which allows the operator to set tasks by means of natural language interface, control, and decision-making system.

Acknowledgements. Researched with the financial support by RFBR grants № 15-07-08309, 15-01-05844.

References

1. Worldwide Commercial Robotics Spending Guide. http://www.idc.com/getdoc.jsp?containerId=IDC_P33201
2. IDC Forecasts Worldwide Spending on Robotics to Reach $135 Billion in 2019 Driven by Strong Spending Growth in Manufacturing and Healthcare. https://www.idc.com/getdoc.jsp?containerId=prUS41046916
3. www.abb.com
4. http://blog.robotiq.com/bid/71048/New-Collaborative-Robot-called-Roberta-Your-New-Best-Friend
5. http://www.bosch-apas.com/en/apas/start/bosch_apas.html
6. http://robot.fanucamerica.com/products/robots/collaborative-robot-fanuc-cr-35ia.aspx
7. http://www.fp-robotics.com/en/
8. Nagoev, Z.V.: Multiagent recursive cognitive architecture. In: Chella, A., Pirrone, R., Sorbello, R., Jóhannsdóttir, K.R. (eds.) Biologically Inspired Cognitive Architectures 2012. AISC, vol. 196, pp. 247–248. Springer, Heidelberg (2013)
9. Nagoev, Z., Nagoeva, O., Tokmakova, D.: System essence of intelligence and multi-agent existential mappings. Hybrid Intell. Syst. **420**, 67–76 (2016). Springer, Switzerland
10. Ivanov, P., Nagoev, Z., Pshenokova, I., Tokmakova, D.: Forming the multi-modal situation context in ambient intelligence systems on the basis of self-organizing cognitive architectures. In: 5th Morld Congress on Information and Communication Technologies (WICT 14–16 December, Morocco) (2015)
11. Anchokov, M.I.: Solution to the problem of training a multi-agent neural network by means of a multi-chromosome genetic algorithm. In: 14th International Conference on Hybrid Intelligent Systems (HIS 2014), Kuwait, pp. 97–103 (2014)
12. Pshenokova, I.A., Kudaev, V.C., Anchokov, M.I., Gladkov, L.A.: Development of software infrastructure based on multi-agent recursive cognitive architecture for objects representation in artificial intelligence systems. Izvestia KBNC RANб **6**(2)(68), 166–173 (2015) (in Russian)
13. Denisenko, V.A., Nagoev, Z.V., Nagoeva, O.V.: Design of computer systems based on cognitive recursive architecture for the problem of synthesis of intellectual behavior agent. Softw. Syst. Comput. Methods **3**, 264–267 (2013). (in Russian)
14. Pshenokova, I.A., Denisenko, V.A., Nagoeva, O.V., Sundukov, Z.A: Knowledge representation in artificial intelligence systems based on the principles of ontoneuromorphogenesis and multi-agent modeling. Izvestia KBNC RAN **6**(2)(68), 158–165 (2015) (in Russian)

Mechanical Leg Design of the Anthropomorphic Robot Antares

Nikita Pavluk[1(✉)], Arseniy Ivin[1], Victor Budkov[1], Andrey Kodyakov[1,2], and Andrey Ronzhin[1,2]

[1] SPIIRAS, 39, 14th line, St. Petersburg 199178, Russia
antei.hasgard@gmail.com
[2] SUAI, 67, Bolshaya Morskaya str., St. Petersburg 190000, Russia

Abstract. An overview of the existing anthropomorphic robots and an analysis of servomechanisms and bearing parts involved in the assembly of robot legs are presented. We propose an option for constructing the legs of the robot Antares under development. A two-motor layout, used in the knee, ensures higher joint power along with independent interaction with the neighboring upper and lower leg joints when bending. To reduce the electrical load on the main battery of the robot, the upper legs are provided with a mounting pad for additional batteries powering servos. Direct control of the servos is also carried out through the sub-controllers, responsible for all 6 engines installed in the articular joints of the robot legs.

Keywords: Anthropomorphic robots · Servomechanisms · Kinematic scheme · Antares · Component parts design · Two-motor knee · Pelvic mechanism

1 Introduction

Among problems, which developers of mobile general-purpose and special-purpose robots face nowadays, are issues of robot's flotation ability on the rugged terrain, autonomous movement and control of kinetic equipment [1]. A lever-hinge system of human and animals mobility, created by nature, is the most adapted to the natural earth's surface and is suitable for use during movement of the anthropomorphic robot [2].

Because of the lack of a unified methodology and software for engineering lever-hinge systems for anthropomorphic robots, developers are forced to create their own software in the design process of each individual robot [3, 4].

The aim of this article is to analyze the existing solutions to the construction of lever-hinge mechanisms of the lower extremities (legs) of anthropomorphic robots and to develop the rational structure of legs of the robot Antares.

One of the simplest structures of the biped robot is described in [5]. It is made up of two-millimeter aluminum sheet, includes six servos operated by EyeBot controller, and weighs 1.11 kg. When walking the robot reaches a speed of 120 m/h at a maximum angle of 60° between the hips. Such robot architecture with six servos is used in [6] to study the operating angles of the joints of the knee, ankle and hip.

© Springer International Publishing Switzerland 2016
A. Ronzhin et al. (Eds.): ICR 2016, LNAI 9812, pp. 113–123, 2016.
DOI: 10.1007/978-3-319-43955-6_15

The anthropomorphic robot in the HanSaRam series, which regularly participates in the FIRA league since 2000, is discussed in [7]. HanSaRam-VIII (HSR-VIII) robot has 28 servos, weighs 5.5 kg and can move at a speed of up to 12 cm/s.

In [8], the anthropomorphic robot Lola, having 7° of freedom per leg, weighs 55 kg with the 180 cm height. The problems of the stability of the robot after stopping as well as a gradual contact of foot parts with the surface when walking are discussed. Elastic materials of the toe and heel of the robot foot ensure reduction of impact force on the robot structure during touching the surface.

Home assistive robot with 14 servos and 16° of freedom, presented in [9], has anthropomorphic upper body architecture and a wheelbase. The paper mostly focuses on two-handed manipulation of objects by robots during transferring domestic objects.

The anthropomorphic robot SWUMANOID, 92 cm high, having in the construction 24 servomechanisms of the series Dynamixel, is designed in [10] to simulate motions and swimming in water. Unstable rolling of robotic body in water complicates the calculation of the kinematics of movements of a floating robot that consists of 21 composite components.

The paper [11] proposes the original software platform for modeling a kinematic motion scheme of anthropomorphic robot legs, where elements of legs are considered as series-connected parts, and recursive algorithms with low computational complexity are used for solving direct and inverse tasks of leg movements.

For moving on complicated uneven surfaces, impassable for tracked or wheeled robots, more sophisticated nonanthropomorphic structures are also being developed, with one [12], four [13], six [14, 15] and a large number [16] of lower extremities.

Based on the conducted analysis of structures of anthropomorphic robots, the robot Poppy of the French company INRIA Flowers and Darwin-OP robot of the company Trossen Robotics were determined to be the closest analogues to the robot Antares under development. Let us consider these robots in more detail.

Details of the anthropomorphic robot Poppy are created based on 3D-printing, which allowed much cheaper production of the robot parts and made it possible to use less powerful servos, but reflected on robot stability when walking. The robot has 25 servomechanisms Dynamixel MX-64 and MX-28 [17], which ensure unobstructed movement of the limbs with a given accuracy and strength margin as gears are made of metal. The robot is controlled by a single-board computer Raspberry Pi and is equipped with 16 force-measuring sensors, 2 HD cameras, a stereo microphone and an inertial measurement unit. Poppy's "face" is an LCD-screen, which can show "emotions" or information about errors.

Modular robot design helps the researcher to change the movement of any robot's limb by isolating the desired limb from the rest of the body, almost without affecting the performance. The structure is specially designed for the installation of additional sensors and connection cables. In addition, such a design facilitates periodic robot maintenance service. However, the center of mass of Poppy is located in the solar plexus, which adversely affects load distribution. The robot becomes unstable and cannot move independently (only with the help of human). Poppy has the same number of degrees of freedom as Antares in the pelvic region, distributing them in a different way, which makes it not so mobile in the knee and ankle joints. The leg joints have only 3 motors (one in each ankle, knee and hip joints). This adversely affects the overall mobility of the robot, judging by the video footage and design files of the robot available from the developers.

The robot Darwin-OP is a robot platform intended for research and development within the framework of educational process. DARwin OP has high performance and dynamic characteristics and a wide range of sensors. The robot communicates with people by using loudspeakers, microphones, cameras, tactile sensors, LEDs, hand gestures. It possesses 20 Dynamixel actuators that ensure free movement of limbs with a given accuracy and strength margin, as the gears are made of metal. The center of mass is located in the center of pelvis, which ensures the correct distribution of the load during walking and inertia, especially in the extremities. A modular robot design helps the researcher to change the movement of any limb of DARwin OP. The structure also allows the installation of additional sensors.

2 Design of Leg Joint of the Anthropomorphic Robot Antares

Designing Antares included several steps associated with the development of joints of legs, arms, torso and head. First priority is given to designing the leg joint because of the following reasons: the high complexity of the layout of joint parts; necessity of this joint for robot movement in space; complexity of calculating joint unit due to the assumed highest load of parts relative to all other joints.

A kinematic scheme (Fig. 1a) displays the overall layout of all the links that make up a single mechanism of robot's lower limbs. A tree graph (Fig. 1b) represents the kinematic structure of the actuating mechanism of the robot. The pelvic mechanism in

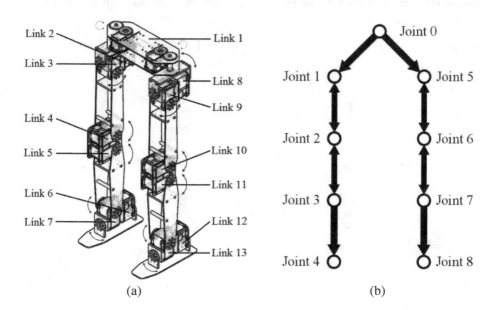

(a)   (b)

Fig. 1. A kinematic scheme and a tree graph of the legs of the Antares robot

both the scheme and the graph is represented by a single joint (Fig. 1b) and a link (Fig. 1a) as it performs only axial rotation function concerning each leg; in the following, actuators of each leg are marked as separate links. Since the kinematic scheme of robot's legs is branching, this causes certain difficulties in its description. Because the sequential arrangement of joints is indicated by links, each of which is responsible for the movement in its plane of movement and is mutually dependent on the closest joints and links. This is because the spatial position of each next link depends on the preceding link in the kinematic scheme, which is also displayed in the graph. The exception is the pelvic mechanism shown at the top of the scheme, since in this case its position in space is the result of the overall mechanism performance. The tree graph (Fig. 1b) helps to understand the relations in the kinematic scheme (Fig. 1a). It shows that the links are combined in joints, except for the end joints 4 and 8, which are the feet of robot's legs. From the graph it is clear that the joints 1, 2, 3 and 5, 6, 7 have bidirectional connections, which is due to the mutual dependence between these joints, as they directly affect the position of each other in space during robot movement.

Total toe structure includes pelvic joint and two identical leg joints consisting of simpler attachment point to the pelvic, hip, knee, lower leg, ankle and foot units. Joints of robot pelvic and legs are developed in accordance with the proportions of the human body, adjusted for assumed growth. Their total length is 510.7 mm. The length of ankle and hip joints is 20 cm. Auxiliary batteries (an installation site is provided) will be installed in the upper legs to reduce the electrical load on the main battery of Antares as well as to control actuator powering, which will help to avoid power supply problems. To save the processing power of the main controller and the computer located in the torso, auxiliary controllers will also be installed in the upper leg joints. These controllers are responsible for the work of all six engines of robotic leg joints. The given design complicates the calculation of the kinematics of the robot motion, but provides more

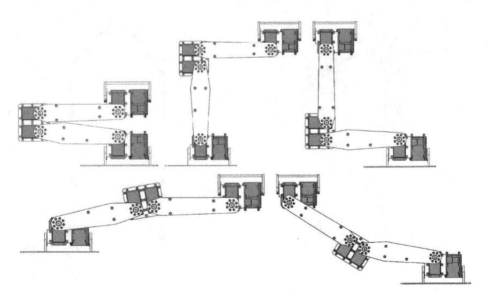

Fig. 2. Functional capabilities of the two-motor knee

complex movements. Through the use of twin-engine layout in the knee a separate leg joint is obtained, which interacts with neighboring upper and lower leg joints, allowing them to be independent of each other when bending (Fig. 2). In addition, the use of two-motor knee simplifies the selection of servo, as in this type of knee the load is divided between two separate motors. Another advantage of this unit is that it facilitates design engineering of the above-mentioned joints of the upper and lower legs.

Interconnected plate lines, linked by screw couplings and cross plates, make up the basis of the leg structure of Antares. For parts manufacturing, 2, 4 and 10 mm thick aluminum sheets were used. 6-mm-thick aluminum rods were applied for the crossties. Parts made from 2 mm thick aluminum sheet are the basis of the joints of lower and upper legs, foot and pelvic attachment. The joints parts made from 2 mm sheet are structurally designed for power loads and pressure from top.

The parts made from 4-mm-thick aluminum sheets are used as stiffening ribs intended for the torsional loads. From these sheets we made transverse upper and lower leg struts and mounting plates of the hip (4 mm thick) and foot. It is also intended to use them to locate the internal components, such as an auxiliary battery, actuators controller of the entire leg joint. Along with the transverse plates, aluminum bars are used as stiffening ribs but only as elements of structural reinforcement.

In the tibial joint a broader transverse plate is used in order to achieve sufficiently reliable structure. This was necessary to ensure that the ankle could be used as efficiently as possible, which required the free use of the internal space of ankle joint. Screw coupling in this case is not only a stiffening rib, but also an arresting stop so that the ankle joint could not be broken and lead to the damage of other components while working. From 10 mm thick aluminum sheets one type of parts is made — a special bearing plate used for the assembly of hip and ankle joints.

For motors Dynamixel MX-64, used in the leg joints, the flanges were made that are necessary for linking separate joints parts and components to provide mobility and stability. The flanges are located in a special socket on the motor housing and the bearing and secured by a special cover which prevents the collapse of the structure during motion and because of vibration. An obligatory requirement for the bearing is a height of 3.5 mm, to comply with the centering of axial arrangement of engines in the overall structure, which is important during robot movement.

Pelvic mechanism is located in the lower part of the torso of Antares and is designed for the axial legs rotations as well as for accommodating the main battery. The construction

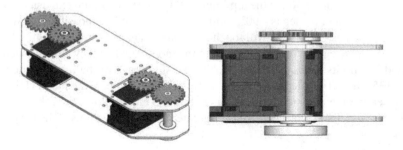

Fig. 3. Pelvic mechanism

includes two bushings, gears with gear ratio 1: 1, two motors Dynamixel MX-28, two 2-mm-thick plates, and the flange bearings 8×14 (Fig. 3). The choice of less powerful servos, compared with those used in the construction of the other legs joints, is substantiated by the fact that for axial rotations the high power of engines is not required.

The hip and ankle joints are formed by pairwise motor connection with metal inserts; in the motor housing extra strong plastic or aluminum are used in order to withstand physical loads on the motor housing during movement. Pairwise motor connection provides space saving in the construction, in order to avoid excessive massiveness. This is necessary to ensure the mobility and flexibility of the assembled joints. In addition, the construction of the hip and ankle joints is designed to the highest possible reproduction of functionality of human ankle and hip joints which are similar to a spherical joint with a limited angle of rotation. Thus, to reproduce the human joint structure, it was decided to introduce to these joints two cylindrical hinges with mutually perpendicular axes. This solution is applied because, at the moment, it is not possible to repeat structure of the spherical joint and make it controllable to the full for the anthropomorphic mechanism, sufficiently reliable and compact, relatively inexpensive and not requiring constant maintenance. Achieving all of this will complicate both the construction itself and the control of the robot. The implemented solution, in turn, simplifies the structure of the robot ankle without compromising ankle capabilities.

3   Experiment Results

The leg construction provides large steering angles for the motors, which ensures greater flexibility and ductility of the joint compared with the mentioned analogs. The robot can easily do the splits, raise the leg straight or at an angle, bend it at the knee, while keeping the foot parallel to the floor, if necessary. There are a variety of sit-ups rather than some specific predetermined movements seen in the analogues, including touching the floor by the pelvic mechanism without the threat of damaging its joints. Pelvic mechanism provides axial rotation of the robot lower limbs, which has a positive effect on its portability and allows it to rotate on the way with a minimum turning radius, turn in the necessary direction on the spot and perform a wider range of different leg movements than those available to humans.

Pelvic mechanics allows the robot to rotate each leg at $360°$ around its axis; however, at this stage we limited it up to $270°$, since there was no need for such a large range of rotation. The angle of flexion at the hip joint is $120°$. The maximum extension angle is $45°$. The side lunge of the leg is $+90°$. Similar to humans, it is impossible to fully bring the leg back, as the lower limbs will touch each other. The maximum bringing back of the leg is $55°$. The legs are brought back up to the moment when they almost touch each other, and this limits the range of the stroke. The side lunge of the leg is limited to an angle of $135°$. It was decided to set limits up to $90°$. This is due to the fact that the robot should resemble the structure of the human, and this angle allows the robot to do the splits. In this case, there is no need to have a greater range of variations of the leg position in the hip joint.

The knee joint comprises two servos combined into one unit. This design solution increases the mobility and strength torque of the knee. This allows the actuator to bend the knee at an angle of 162°. The use of two actuators in the knee joint entails constant synchronization of the motors relative to each other, since one actuator does not provide the same large angle of the displacement in a joint without a significant loss in power and mobility of the structure. The result is that the location of motors corresponds to the perpendicular position relative to the frontal plane of the robot. The ankle design provides the eversion — the displacement of the foot inwards to the sagittal plane as well as the inversion — shifting of the foot outwards from the sagittal plane. Changing the generalized coordinate of the ankle joint in roll represents a rotation of the foot relative to the upper limb from the neutral position. The angle of the foot position changes: eversion is 90° and inversion is 90°. Moving toe upwards relative to the neutral position is taken as a positive angle, and downwards — as the negative angle. In this case, the pitch angle of the foot deviates from +88° to −180°. Figure 4 shows the results of work to change the generalized coordinate of the robot foot. Movement is carried out in full compliance with design feature that limits the angles of pitch and roll.

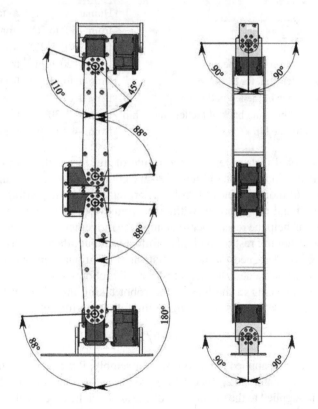

Fig. 4. Turning angles of the leg

Apart from the pelvic mechanism responsible for axial rotations of legs and being a separate unit, in the legs structure we applied actuators Dynamixel MX-64. Each

Dynamixel has a unique ID for connection to the common data line, supports the connection TTL, RS485 and others, can be connected to the common control bus; available LED or emergency shutdown (torque-off) functions may be set to predetermined values of temperature, current and voltage. These actuators can be adjusted to move more smoothly. Dynamixel servos can be controlled from a PC or a microcontroller, which is a great advantage in the development of prototypes. Without the battery and the controller installed in the hip joint, the estimated total weight of the metal frame, servos and flange connections is 1.07 kg, of which 756 g account for the Dynamixel MX-64 servos. Construction of the two joints of the leg and pelvic mechanism weighs 2.44 kg.

Table 1 shows a comparison of the different angles of deflection of legs positions of the human and developed prototype of an anthropomorphic robot Antares. Based on this table, we can conclude that the various actions typical of the human and the robot can be executed by the latter, as the ranges of changes of human positions angles lie within the ranges of changes of robot angles.

Table 1. The comparison of the angles of the human and the robot Antares

	Human	Antares
Range by the pitch of foot	from −50° to 40°	from −88° to 180°
Range by the roll of foot	from −30° to 20°	from −90° to 90°
Range of flexion-extension, for the hip	from +120° to 20°	from +110° to 45°
Range of flexion-extension, for the knee	from 0° to 110°	from 0° to 162°
Range of the side lunge of the leg in the hip	from 0° to 45°	from 0° to 90°
Range of the relative bringing back of the leg in the hip	from 0° to 30°	from 0° to 55°
Range of rotation for the hip	from −45° to 45°	from −45° to 45°

It should be noted that, except for the range of the foot roll, all other angles are artificially limited in order to avoid unnecessary contact of the parts lying in a single plane at the time of motion. Since the robot represents a robotic platform intended for scientific research and development within the framework of educational process, its modular nature can help the researcher to change the movement of any limb by isolating the desired limb from the rest of the body, almost without affecting performance. The structure is specially designed for the installation of additional sensors and connection cables. In addition, such a design facilitates periodic robot maintenance service.

Load weight balancing in the legs of the robot taking into account the maximum supposed weight of 8 kg occurs according to the following formula:

$$F = P/2 = 80/2 = 40N,$$

where, F is the force applied to the lower leg assembly; P is a weight, which keeps the lower leg; the denominator — 2, since all the weight is distributed over 2 legs. Figure 5 shows the torque supplied to the motor shaft of the ankle in the process of raising the legs. The graph shows that the ankle motor has a maximum torque when the robot is in a sitting position. The tension in the construction of the lower and upper leg with a total weight of the robot equal to 8 kg is $7.44 \cdot 10^5$ N/m^2. The conducted study on the maximum load revealed that the lower leg can withstand a load of 1040 Newtons. Summarizing, we can say that the lower and upper leg individually can withstand 13-fold vertical load on the structure, without succumbing to deformation and breakage. It was taken into consideration that fasteners for flange connections were fixed and immobile.

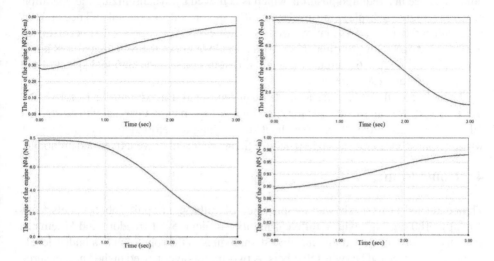

Fig. 5. The dependence of the torque of the motors № 2 to № 5 on their operational time

The graphs show that, except for the motor № 2, the maximum motor torque appears to be at the beginning of the time interval and decreases with time. The graph in Fig. 5 shows that the ankle motor has the maximum torque when the robot is in a standing position, as keeping this position is the most time-consuming task for this type of engine. For the motor № 2, the shape of the curve remains practically unchanged with increasing or decreasing the load. With the motors № 3 and № 4 a similar situation may be observed. The graph № 2 practically mirrors the graph № 4, while the graph № 3 is almost identical to the graph № 5 because the program in simulation grouped the motors. The first group includes the motors № 2 and № 3, the second – motors № 4 and № 5. In a group one

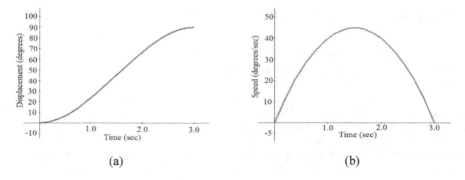

(a)   (b)

Fig. 6. (a) displacement of shafts of the motors № 2 to № 5 and their operational time; (b) rotational speed of shafts of the motors № 2 to № 5 and their operational time

motor is responsible for the leg rise, and the other for holding the knee and hip in a position perpendicular to the support surface. In this simulation the motors № 2, 3 and 4 are used to lift the robot leg. Apart from performing leg lifting, the motor № 3 also holds the knee in a required position, which is achieved by synchronizing the operation of motors. Figure 6 displays the displacement of motor shafts of the ankle, knee and hip joints when the robot starts to move from a standing position. The tension occurring in the structure of the lower or upper leg, with a total weight of the robot equal to 8 kg, is $7.44 \cdot 10^5$ N/m^2. Figure 6 presents the angular displacement and rotational speed of the shafts of the motors № 2 to № 5.

According to the study, the lower leg can withstand a maximum load of 1040 N. Summarizing, we can say that the lower and upper legs individually can withstand a 13-fold vertical load on the structure, without succumbing to deformation and breakage. It was taken into consideration that fasteners for flange connections were fixed and immobile.

4   Conclusion

The conducted analysis showed the presence of anthropomorphic robot models from 30 cm to 180 cm high with a different number of degrees of freedom and kinematic schemes. It was concluded that the closest analogues to the robot Antares under development are Poppy and Darwin-OP robots. A two-motor layout, used in the knee, ensures higher joint power along with independent interaction with the neighboring upper and lower leg joints when bending. The electrical load on the main battery of Antares is reduced by the use of auxiliary batteries installed in the upper legs and powering the servos. The direct servo control is also performed by the auxiliary controllers responsible for the work of all six engines of the leg joints. Studies of the prototype design have demonstrated that individual components and parts have a more than ten-fold safety margin. The robot under development is intended for development of assistive technologies of human-computer interaction based on multimodal interfaces and use for educational purposes such as participation in robot football competitions [18–24].

Acknowledgment.   The study was performed through the grant of the Russian Science Foundation (project №16-19-00044).

References

1. Kudryashov, V.B., Lapshov, V.S., Noskov, V.P., Rubtsov, I.V.: Problems of robotization for military ground technics. Izvestiya SFedU. Eng. Sci. **3**(152), 42–57 (2014)
2. Kovalchuk, A.K., Kulakov, D.B., Semenov, S.E., Yarots, V.V., Vereikin, A.A., Kulakov, B.B., Karginov, L.A.: Method for designing spatial tree-like actuators of walking robots. Eng. Bull. Bauman MSTU **11**, 6–10 (2014)
3. Karpenko, A.P.: Robotics and computer-aided design systems. Teaching guide. MGTU im. N.E. Baumana, Moscow (2014)
4. Zeltser, A.G., Vereikin, A.A., Goyhman, A.V., Savchenko, A.G., Zhukov, A.A., Demchenko, M.A.: The concept of capsular exoskeleton for rescue operations. Eng. Bull. Bauman MSTU **3**, 14–22 (2015)

5. Warnakulasooriyaa, S., Bagheria, A., Sherburnb, N., Shanmugavel, M.: Bipedal walking robot – a developmental design. Procedia Eng. **41**, 1016–1021 (2012)
6. Lima, S.C., Yeapa, G.H.: The locomotion of bipedal walking robot with six degree of freedom. Procedia Eng. **41**, 8–14 (2012)
7. Yoo, J.K., Lee, B.J., Kim, J.H.: Recent progress and development of the humanoid robot Hansaram. Rob. Auton. Syst. **57**, 973–981 (2009)
8. Buschmann, T., Lohmeier, S., Ulbrich, H.: Humanoid robot Lola: design and walking control. J. Physiol. **103**, 141–148 (2009)
9. Mohameda, Z., Capi, G.: Development of a new mobile humanoid robot for assisting elderly people. Procedia Eng. **41**, 345–351 (2012)
10. Nakashima, M., Tsunoda, Y.: Improvement of crawl stroke for the swimming humanoid robot to establish an experimental platform for swimming research. Procedia Eng. **112**, 517–521 (2015)
11. Shah, S.V., Saha, S.K., Dutt, J.K.: Modular framework for dynamic modeling and analyses of legged robots. Mech. Mach. Theory **49**, 234–255 (2012)
12. Yua, X., Fub, C., Chen, K.: Modeling and control of a single-legged robot. Procedia Eng. **24**, 788–792 (2011)
13. Potts, A.S., Jaime da Cruz, J.: A comparison between free motion planning algorithms applied to a quadruped robot leg. IFAC-Papersonline **48**(19), 019–024 (2015)
14. Rostro-Gonzalez, H., Cerna-Garcia, P.A., Trejo-Caballero, G., Garcia-Capulin, C.H., Ibarra-Manzano, M.A., Avina-Cervantes, J.G., Torres-Huitzil, C.: A CPG system based on spiking neurons for hexapod robot locomotion. Neurocomputing **170**, 47–54 (2015)
15. Pan, P.S., Wu, C.M.: Design of a hexapod robot with a servo control and a man-machine interface. Rob. Comput.-Integr. Manuf. **28**, 351–358 (2012)
16. Vidoni, R., Gasparetto, A.: Efficient force distribution and leg posture for a bio-inspired spider robot. Rob. Auton. Syst. **59**, 142–150 (2011)
17. ROBOTIS catalogue. http://en.robotis.com/index/product.php?cate_code=101011. Accessed 3 Mar 2016
18. Motienko, A.I., Makeev, S.M., Basov, O.O.: Analysis and modeling of position choice process for transportation of the sufferer on the basis of Bayesian belief networks. SPIIRAS Proc. **43**, 135–155 (2015)
19. Ronzhin, A., Budkov, V.Y.: Multimodal interaction with intelligent meeting room facilities from inside and outside. In: Balandin, S., Moltchanov, D., Koucheryavy, Y. (eds.) NEW2AN/ruSMART 2009. LNCS, vol. 5764, pp. 77–88. Springer, Heidelberg (2009)
20. Ronzhin, A.L., Budkov, V.Y., Ronzhin, A.L.: User profile forming based on audiovisual situation analysis in smart meeting room. SPIIRAS Proc. **23**, 482–494 (2012)
21. Ronzhin, A.L., Karpov, A.A., Leontyeva, A.B., Kostuchenko, B.E.: The development of the multimodal information kiosk. SPIIRAS Proc. **5**, 227–245 (2007)
22. Suranova, D.A., Meshcheryakov, R.V.: Personified voice interaction software in billing systems. In: Ronzhin, A., Potapova, R., Delic, V. (eds.) SPECOM 2014. LNCS, vol. 8773, pp. 345–352. Springer, Heidelberg (2014)
23. Karpov, A.A., Ronzhin, A.L.: Information enquiry kiosk with multimodal user interface. Pattern Recogn. Image Anal. **19**(3), 546–558 (2009). MAIK Nauka/Interperiodica
24. Yusupov, R.M., Ronzhin, A.L.: From smart devices to smart space. Herald Russ. Acad. Sci. MAIK Nauka **80**(1), 45–51 (2010)

Method of Spheres for Solving 3D Formation Task in a Group of Quadrotors

Donat Ivanov$^{(\boxtimes)}$, Sergey Kapustyan, and Igor Kalyaev

Scientific Research Institute of Multiprocessor Computer Systems
of Southern Federal University, Taganrog, Russia
donat.ivanov@gmail.com

Abstract. Formation task in a group of unmanned aerial vehicles (UAVs) is a very important problem in a multi-robotics. The paper contains a brief analysis of existing methods for formation task in groups of robots. A new method for solving formation task in a group of quadrotors is proposed. The method make it possible to ensure accurate compliance with distances between quadrotors in the formation, as well as featuring low computational complexity.

Keywords: Formation task · Milti-robotics · Quadrotor · Unmanned aerial vehicle

1 Introduction

The progress in a microelectronics and a computing make it possible to produce small-sized UAV, which can be cheap and easily accessible in case of mass production. However, practical possibilities of a single UAV are limited. As widely shown in the recent literature [1], robustness and flexibility constitute the main advantages of multiple-robot systems vs. single-robot ones. Also the use of group of UAVs opens wide perspectives for unmanned aircraft [2, 3].

Micro Aerial Vehicle (MAV) can be classified on: airplane, helicopter, bird-like, insect-like, autogiro and blimp. The work [4] compare a lot of characteristics like power cost, control cost, payload/volume, maneuverability, mechanics simplicity, aerodynamics complexity, stationary flight, low speed fly, high speed flight, survivability, vulnerability, vertical takeoff and landing, endurance, miniaturization, indoor usage. There are a lot of practical applications where some of the characteristics are much important then some others, which make it possible and reasonable to use other types of small-sized UAVs. There is a comparison of various types of MAV in [4], where showed that quadrotors are the most universal type of MAV. It does not mean that other types of small-sides UAVs are not important and useful. All of them could be useful in various specific spheres and applications.

Groups of unmanned quadrotors can be used for a video monitoring, forming phased antenna arrays and other applications (see for example [5] and the references therein). This applications requires mutual position relative to each other quadrotors in a group. The target location of quadrotors in space is named "formation", and the task of forming the formation is named "formation task".

© Springer International Publishing Switzerland 2016
A. Ronzhin et al. (Eds.): ICR 2016, LNAI 9812, pp. 124–132, 2016.
DOI: 10.1007/978-3-319-43955-6_16

2   Known Methods for Solving Formation Task

There are some well-known methods for solving a formation task in a group of mobile robots, including a behavior based [6, 7], leader-follower approach [8–12], a virtual structure/virtual leader approach [13], based on the game theory [14] etc. However some of the methods make it possible to form a formation by only a certain set of shapes. Some others require considerable computing resources, but there are not such resources on-board of quadrotors.

Many of the methods are aimed at positioning quadrotors in absolute coordinates, but in practice compliance required distances between the quadrotor is more important than the quadrotor's positioning in absolute coordinates.

Thus the formation task in a group of quadrotors needs a computationally simple method, which provide to derive desired target formation of various sharps and precise distances between quadrotors. Such method for solving 2D formation task on a plane is proposed in [15]. This work is continuing this approach to 3D formations.

3   Formal Statement of the Formation Task in a Group of Quadrotors

There is a set $\mathbf{R}$ of controlled quadrotors $r_i \in \mathbf{R}(i = \overline{1,N})$, where N – is the number of quadrotors in a group. The status of each quadrotor $r_i \in R$ is described by the vector of status $\mathbf{s}_i(t) = [s_{i,1}(t), s_{i,2}(t), \ldots, s_{i,h}(t)]^T$, where the state variable $s_{i,h}(t)$ mean quadrotor's coordinates $x_i(t), y_i(t), z_i(t)$, current speed, acceleration, roll $\varphi_i(t)$, pitch $\theta_i(t)$ and yaw $\psi_i(t)$ angles, the remaining board energy reserves, etc.

Current mutual arrangement of quadrotors in a group is described by the matrix

$$\mathbf{D}_t = \begin{bmatrix} 0 & d_{1,2}(t) & d_{1,3}(t) & \cdots & d_{1,N}(t) \\ - & 0 & d_{2,3}(t) & \cdots & d_{2,N}(t) \\ - & - & 0 & \ddots & \vdots \\ - & - & - & 0 & d_{N-1,N}(t) \\ - & - & - & - & 0 \end{bmatrix},$$

where each element $d_{i,j}(t)$ of matrix $\mathbf{D}(t)$ represents the distance between quadrotors r_i and r_j the current time.

Each quadrotor $r_i \in \mathbf{R}$ has information about their own condition $\mathbf{s}_i(t)$, and information about distances $d_{i,j}(t)(i,j = \overline{1,N}, i \neq j)$ between the quadrotor $r_i \in \mathbf{R}$ and other quadrotors $r_j \in \mathbf{R}(j = \overline{1,N}, j \neq i)$. Each quadrotor $r_i \in \mathbf{R}$ has a control system, which make it possible to change coordinates $\mathbf{x}_i(t) = x_i(t), y_i(t), z_i(t)$ according to control inputs $u_i(t)$ based on the mathematical model considered in [4].

In order to prevent collisions and a mutual interference of quadrotors, the limit of positions is introduced:

$$\left| \mathbf{x}_i(t) - \mathbf{x}_j(t) \right| \geq \Delta_r, (i \neq j; i,j = \overline{1,N}), \tag{1}$$

where Δ_r – is a minimal acceptable distance between quadrotors, excludes mutual interference of quadrotors.

The target formation is a set $\mathbf{V}$ of target positions $v_\mu \in \mathbf{V}(\mu = \overline{1,N})$ of single quadrotors. Each target position $v_\mu \in \mathbf{V}$ is described by a point $p_\mu(\mu = \overline{1,N})$ with coordinates (x_μ, y_μ, z_μ). But there is not information about point's $p_\mu(\mu = \overline{1,N})$ coordinates $x_{p\mu}, y_{p\mu}, z_{p\mu}(\mu = \overline{1,N})$ and about assignments between quadrotors $r_i \in \mathbf{R}$ and target positions $v_\mu \in \mathbf{V}$.

The only one available information about the target formation is a matrix

$$\mathbf{D}_f = \begin{bmatrix} 0 & d_{1,2} & d_{1,3} & \cdots & & d_{1,N} \\ - & 0 & d_{2,3} & \cdots & & d_{2,N} \\ - & - & 0 & \ddots & & \vdots \\ - & - & - & 0 & d_{N-1,N} \\ - & - & - & - & 0 \end{bmatrix},$$

where each variable $d_{i,j}$ of $\mathbf{D}_f$ is a distance between points p_i and p_j of target positions v_i and v_j in a target formation.

Formation task in the group of quadrotors is to determine a sequence of controls $\mathbf{u}(t) = [u_1(t), u_2(t), \ldots u_N(t)]^T$ which lead a group from start formation with distances $\mathbf{D}(t_0)$ to desired target formation with distances $\mathbf{D}_f$ for minimum time and with a restrictions on the quadrotor's positions (1).

4 The Proposed Method

We propose the following method for solving formation task, which is named "methods of spheres", which is continuing the "method of circles" [15].

At the first step we need to choose the target position v_μ and the quadrotor r_i, which are used for the beginning of formation building. Point p_c is the center of the group. The point p_c is described by radius vector $\overrightarrow{l_c}$:

$$\overrightarrow{l_c} = \frac{1}{N} \sum_{i=1}^{N} \overrightarrow{l_i},$$

where $\overrightarrow{l_i}$ is a radius vector of a quadrotor $r_i \in \mathbf{R}(i = \overline{1,N})$ positions $x_i(t), y_i(t), z_i(t)$. After that distances between point p_c and point of r_i current position should be calculated:

$$l_{i,c} = \sqrt{(x_i - x_{c_1})^2 + (y_i - y_{c_1})^2 + (z_i - z_c)^2}.$$

Then find the quadrotor with a minimal distance $\min(l_{i,c}), i \in \overline{[1,N]}$ between itself and the center of the group.

The nearest for the point p_c quadrotor r_i get an assignment with target position v_1. And the current coordinates of the quadrotor r_i is a coordinates p_1 of the target position v_1.

At the second step the assignment for target position v_2 and its location $p_2(x_{p2}, y_{p2}, z_{p2})$ is determined. For this construct a sphere $c_{1,2}$ with center in point p_1 and radius $d_{1,2}$ (from the matrix $\mathbf{D}_f$). Then construct $N - 1$ straight lines, each one passing through the point p_1 and the current position of quadrotors $r_i (i = \overline{2,N})$

$$\frac{x_{p2i} - x_1}{x_i - x_1} = \frac{y_{p2i} - y_1}{y_i - y_1} = \frac{z_{p2i} - z_1}{z_i - z_1}, \quad i \in \overline{[2,N]}.$$

To determine the coordinates of the intersection points of these lines and the sphere $c_{1,2}$ for each quadrotor $r_i (i = \overline{2,N})$ use the system of equations:

$$\begin{cases} \dfrac{x_{p2i} - x_1}{x_i - x_1} = \dfrac{y_{p2i} - y_1}{y_i - y_1}; \\ \dfrac{x_{p2i} - x_1}{x_i - x_1} = \dfrac{z_{p2i} - z_1}{z_i - z_1}; \qquad i \in \overline{[2,N]} \\ (x_{p2i} - x_1)^2 + (y_{p2i} - y_1)^2 + (z_{p2i} - z_1)^2 = d_{1,2}^2. \end{cases}$$

and get equation's roots $(x_{p2i_1}, y_{p2i_1}, z_{p2i_1})$ and $(x_{p2i_2}, y_{p2i_2}, z_{p2i_2})$ for each $r_i (i = \overline{2,N})$. Then calculate distances between quadrotors $r_i (i = \overline{2,N})$ and their points $(x_{p2i_1}, y_{p2i_1}, z_{p2i_1})$ and $(x_{p2i_2}, y_{p2i_2}, z_{p2i_2})$:

$$l_{i,p2iq} = \sqrt{(x_i - x_{p2i_q})^2 + (y_i - y_{p2i_q})^2 + (z_i - z_{p2i_q})^2}, \quad q \in \overline{[1;2]}, \ i \in \overline{[2,N]}.$$

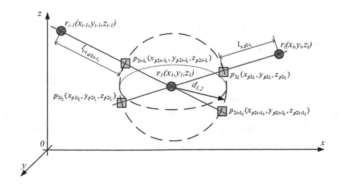

Fig. 1. Definition of the target position's point p_2

The quadrotor with the minimal length of distance $\min(l_{i,p2iq}), q \in \overline{[1,2]} i \in \overline{[2,N]}$ get an assignment target position v_2 and point p_2 with coordinates (x_{p2}, y_{p2}, z_{p2}), like it is showed at Fig. 1.

At the third step the coordinates $p_3 (x_{p3}, y_{p3}, z_{p3})$ of the target position v_3 is determined. For this construct two spheres $c_{1,3}$ and $c_{2,3}$. The sphere $c_{1,3}$ has the center in point p_1 and the radius $d_{1,3}$. The sphere $c_{2,3}$ has the center in point p_2 and the radius $d_{2,3}$. Then find the set of points of intersections of spheres $c_{1,3}$ and $c_{2,3}$. It is a circle:

$$\begin{cases} (x_{p3} - x_1)^2 + (y_{p3} - y_1)^2 + (z_{p3} - z_1)^2 = d_{1,3}^2; \\ (x_{p3} - x_2)^2 + (y_{p3} - y_2)^2 + (z_{p3} - z_2)^2 = d_{2,3}^2. \end{cases}$$

Then find points $p_{3i}, i \in \overline{[3,N]}$, which are nearest for quadrotors $r_i (i = \overline{3,N})$. Then calculate distances $l_{3,i}[i = \overline{3,N}]$ between quadrotors $r_i (i = \overline{3,N})$ and each of roots $(x_{p3,q}, y_{p3,q}), q \in \overline{[1,2]}$:

$$l_{p3q} = \sqrt{(x_i - x_{p3_q})^2 + (y_i - y_{p3_q})^2 + (z_i - z_{p3_q})^2}, \quad i \in \overline{[3,N]}, q \in \overline{[1,2]}.$$

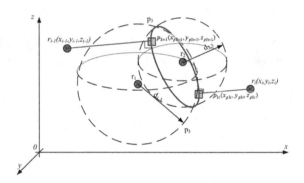

Fig. 2. Definition of the point $p_3 (x_{p3}, y_{p3}, z_{p3})$ of target position v_3

The quadrotor with the minimal length of distance $\min(l_{p3q}), i \in \overline{[3,N]}, q \in \overline{[1,2]}$ get an assignment target position v_3 and point p_3 with coordinates (x_{p3}, y_{p3}, z_{p3}), like it is showed at Fig. 2.

At the fourth step the coordinates $p_4 (x_{p4}, y_{p4}, z_{p4})$ of the target position v_4 is determined. For this construct three spheres $c_{1,4}$, $c_{2,4}$ and $c_{3,4}$. The sphere $c_{1,4}$ has the center in point p_1 and the radius $d_{1,4}$. The sphere $c_{2,4}$ has the center in point p_2 and the radius $d_{2,4}$. The sphere $c_{3,4}$ has the center in point p_3 and the radius $d_{3,4}$. Then find points of intersections of spheres $c_{1,4}$, $c_{2,4}$ and $c_{3,4}$:

$$\begin{cases} \left(x_{p4} - x_1\right)^2 + \left(y_{p4} - y_1\right)^2 + \left(z_{p4} - z_1\right)^2 = d_{1,4}^2; \\ \left(x_{p4} - x_2\right)^2 + \left(y_{p4} - y_2\right)^2 + \left(z_{p4} - z_2\right)^2 = d_{2,4}^2; \\ \left(x_{p4} - x_3\right)^2 + \left(y_{p4} - y_3\right)^2 + \left(z_{p4} - z_3\right)^2 = d_{3,4}^2. \end{cases}$$

At the each next step the coordinates $p_\mu \left(\mu = \overline{4, N}\right)$ of target position v_μ is determined. For this construct 4 spheres $c_{\mu-4,\mu}$, $c_{\mu-3,\mu}$, $c_{\mu-2,\mu}$ and $c_{\mu-1,\mu}$ with centers at $p_{\mu-4}$, $p_{\mu-3}$, $p_{\mu-2}$, $p_{\mu-1}$, and radiuses $d_{\mu-4,\mu}$, $d_{\mu-3,\mu}$, $d_{\mu-2,\mu}$, $d_{\mu-1,\mu}$.

Then find points of intersections of spheres $c_{\mu-4,\mu}$, $c_{\mu-3,\mu}$, $c_{\mu-2,\mu}$ and $c_{\mu-1,\mu}$, use the system of equations:

$$\begin{cases} \left(x_{p\mu} - x_{\mu-4}\right)^2 + \left(y_{p\mu} - y_{\mu-4}\right)^2 + \left(z_{p\mu} - z_{\mu-4}\right)^2 = d_{\mu-4,\mu}^2; \\ \left(x_{p\mu} - x_{\mu-3}\right)^2 + \left(y_{p\mu} - y_{\mu-3}\right)^2 + \left(z_{p\mu} - z_{\mu-3}\right)^2 = d_{\mu-3,\mu}^2; \\ \left(x_{p\mu} - x_{\mu-2}\right)^2 + \left(y_{p\mu} - y_{\mu-2}\right)^2 + \left(z_{p\mu} - z_{\mu-2}\right)^2 = d_{\mu-2,\mu}^2; \\ \left(x_{p\mu} - x_{\mu-1}\right)^2 + \left(y_{p\mu} - y_{\mu-1}\right)^2 + \left(z_{p\mu} - z_{\mu-1}\right)^2 = d_{\mu-1,\mu}^2. \end{cases}$$

Then calculate distances $l_{i,p\mu iq}\left[i = \overline{\mu, N}\right]$ between quadrotors $r_i(i = \overline{k, N})$ and each of roots $\left(x_{p\mu q}, y_{p\mu q}, z_{p\mu q}\right)$:

$$l_{i,p\mu iq} = \sqrt{\left(x_i - x_{p\mu i_q}\right)^2 + \left(y_i - y_{p\mu i_q}\right)^2 + \left(z_i - z_{p\mu i_q}\right)^2}, \; q \in \overline{[1; 2]}, i \in \overline{[\mu, N]}.$$

The quadrotor with the minimal length of the distance $\min(l_{i,p\mu iq}), i \in \overline{[\mu, N]}, q \in \overline{[1, 2]}$ get an assignment target position v_μ and the point p_μ with coordinates $\left(x_{p\mu}, y_{p\mu}, z_{p\mu}\right)$.

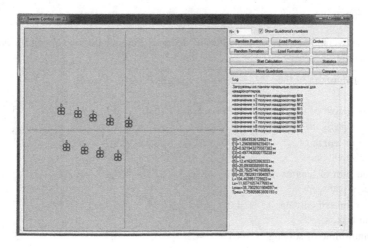

Fig. 3. Computer modeling software

When all assignments and coordinates of all target positions are obtained, each quadrotor moves to its own target position by straight line, except the case with threat of collisions between quadrotors. In latter case quadrotors circumnavigates each other along the arc.

5   Computer Modeling and Experiments

The proposed approach to derive desired target formation was tested by using the software model (Fig. 3) and the experimental stand, using a group of quadrotors.

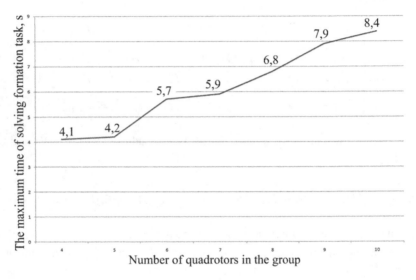

Fig. 4. The maximum time of solving formation task for different numbers of quadrotors within a group, s

Table 1. The average and maximum values of distance and time during solving formation task using the method of spheres

Number of quadrotors in the group	4	5	6	7	8	9	10
The average length of quadrotor's paths, m	8,25	6,8	10,4	9,1	12,7	11,5	13,8
The maximum length of quadrotor' paths, m	20,0	20,1	28,5	28,7	33,6	39,1	41,9
The average sum of lengths of quadrotor's paths, m	33,0	33,9	62,7	64,1	102,2	104,5	112,4
The average time of solving formation task, s	4,0	4,1	5,6	5,7	6,7	7,7	8,3
The maximum time of solving formation task, s	4,1	4,2	5,7	5,9	6,8	7,9	8,4

An information exchange between quadrotors into a group and with the operator's control panel implemented by Wi-Fi. Inertial navigation system is used. Formation task was solved on a horizontal plane, when all quadrotors into a group fly on at the same altitude (~ 1 m). The average horizontal velocity of quadrotors is 5 m/s. The average and maximum values of distance and time during solving formation task using the method of spheres is shown in the Table 1 and Fig. 4.

In contrast to the well-known quadrotors formation projects [16], in that experiments external cameras or sensors were nor used. Quadrotors determined its positions only by on-board sensors.

6 Conclusions

In this paper we consider the problem of formation task on the plane for those cases where the mutual position of quadrotors is more important rather than absolute coordinates of their positions in space.

The algorithm based on a proposed method has a low computational complexity, allows to create formations with various shapes, and opens up opportunities for practical application groups of quadrotor UAV for video monitoring, forming phased antenna arrays and mobile telecommunication systems.

Acknowledgement. This work was supported in part bu the Russian Foundation for Basic Research under Grants № 14-08-01533, № 16-08-00875-a, № 16-58-00226.

References

1. Franchi, A., Secchi, C., Ryll, M., Bulthoff, H.H., Giordano, P.R.: Shared control: balancing autonomy and human assistance with a group of quadrotor UAVs. IEEE Robot. Autom. Mag. **19**(3), 57–68 (2012)
2. Schwager, M., Julian, B., Angermann, M., Rus, D.: Eyes in the sky: decentralized control for the deployment of robotic camera networks. Proc. IEEE **99**(9), 1541–1561 (2011)
3. Fink, J., Michael, N., Kim, S., Kumar, V.: Planning and control for cooperative manipulation and transportation with aerial robots. Int. J. Robot. Res. **30**(3), 324–334 (2010)
4. Bouabdallah, S.: Design and control of quadrotors with application to autonomous flying. Lausanne Polytechnic University (2007)
5. Tonetti, S., Hehn, M., Lupashin, S., D'Andrea, R.: Distributed control of antenna array with formation of UAVs. World Congress **18**(1), 7848–7853 (2011)
6. Balch, T., Arkin, R.C.: Behavior-based formation control for multirobot teams. IEEE Trans. Robot. Autom. **6**, 926–939 (1998)
7. Jonathan, R., Lawton, T., Randal, W., Beard, B.J.Y.: A decentralized approach to formation maneuvers. IEEE Trans. Robot. Autom. **19**(6), 933–941 (2003)
8. Wang, P.K.C.: Navigation strategies for multiple autonomous mobile robots moving in formation. J. Robot. Syst. **8**(2), 177–195 (1991)
9. Desai, J.P., Ostrowski, J., Kumar, V.: Controlling formations of multiple mobile robots. In: Proceedings of IEEE International Conference on Robotics and Automation, pp. 2864–2869 (1998)

10. Mesbahi, M., Hadaegh, F.Y.: Formation flying control of multiple spacecraft via graphs, matrix inequalities, and switching. AIAA J. Guidance Control Dynam. **24**(2), 369–377 (2000)
11. Wang, P.K.C., Hadaegh, F.Y.: Coordination and control of multiple microspacecraft moving in formation. J. Astronaut. Sci. **44**(3), 315–355 (1996)
12. Desai, J., Ostrowski, J., Kumar, V.: Control of formations for multiple robots. In: Proceedings of the IEEE International Conference on Robotics and Automation, Leuven, Belgium (1998)
13. Lewis, M.A., Tan, K.H.: High precision formation control of mobile robots using virtual structures. Auton. Robot. **4**, 387–403 (1997)
14. Erdoğan, M.E., Innocenti, M., Pollini, L.: Obstacle avoidance for a game theoretically controlled formation of unmanned vehicles. In: 18th IFAC, pp. 6023–6028 (2011)
15. Ivanov, D., Kalyaev, I., Kapustyan, S.: Method of circles for solving formation task in a group of quadrotor UAVs. In: 2nd International Conference on Systems and Informatics (ICSAI), pp. 236–240. IEEE (2014)
16. Flying Machine Arena. http://www.flyingmachinearena.org/

Mimic Recognition and Reproduction in Bilateral Human-Robot Speech Communication

Arkady Yuschenko, Sergey Vorotnikov, Dmitry Konyshev[✉], and Andrey Zhonin

BMSTU Robototechnika, Izmayloovskaya sq., 7,
Moscow 105037, Russian Federation
konyshev-dmitri@yandex.ru

Abstract. This paper describes the principles of mimic recognition and reproduction for robots with ability to simulate facial emotional states of a human face. A new method of describing and managing the emotional state from basic emotional states is proposed. The method introduces a more natural way to control facial states and can be combined with a proposed natural speech interface model. The research results were tested in real conditions on several events using robots with silicone human-like faces.

Keywords: Service robotics · Emotional state · Mimic recognition · Mimic reproduction · Facial actuators · Natural interface · Speech interface

1 Introduction

One the most quickly developing niches of contemporary robotics is service robotics, that is oriented on direct interaction between humans and robots, whereas human is not specially trained for that interaction. Development of service robotics gave some results in natural (or close to natural) speech interfaces. Control task for human-operator transforms into verbal dialogue between operator and robot. The latter may inform the operator of the current situation and ask to define more precisely his commands. Often the speech dialogue is not sufficient to "mutual understanding". In such cases may be applied systems that use gestures, which is important for people with hearing and speech dysfunctions. Lately new results were acquired in the field of human-like robots with skin imitation. Such robots allow to reproduce the mimic of a speaking person [1, 2]. With mimics the robot-companion can express its attitude to perceived messages, own current state and environment. Whereas, unlike speech, the mimic facial state does not need to be translated. Mimics represent a common image of evaluation in accordance with behavioral rules and goals, provided with robot's knowledge base. However, more complicated problems take their place. In common, there are two interrelated problems of control: the first is how to describe the intellectual components of processes, the second is about the executive part of robot's mimic apparatus. Another problem is the speech and mimic correlation. Here the main problem is to recognize the sense of the verbal information. Using the mimic support we suppose to raise the effectiveness of human-robot speech communication.

© Springer International Publishing Switzerland 2016
A. Ronzhin et al. (Eds.): ICR 2016, LNAI 9812, pp. 133–142, 2016.
DOI: 10.1007/978-3-319-43955-6_17

2   The Problem of Mimic Recognition

First group of problems presumes psychological analysis and emotion formalization. Fominykh [3, 4] attempted in his research to describe emotions as a result of internal reflection, that is internal evaluation of messages of one speaker in terms of own interests. Author proposed "emotional algebra", which allows calculating characteristics of emotional reaction on a set of given situations. Perspectives of using emotions in human-like robotics and some approaches of implementation are described in works of Ishiguro, Scholtz and others [5–7].

First of we need to identify basic emotional states (BES), that can be recognized by people independent of their nationality and cultural level. P. Ekman proposed 6 BES based on his psychological works [8, 9]. They are happiness, sadness, anger, fear, surprise and disgust. They should be expanded with 7th BES – the neutral state. So, the problem is lowered down to imitate the 7 mimic states.

While describing facial expressions of human face Ekman has developed FACS – a facial expression coding system. FACS declared a facial state in terms of typical action units (AU). AUs describe movement of distinctive facial points, which most often depict movement of a separate facial muscle. The deficiency of this system is the binary nature of AUs (e.g. "left eyebrow raised") and do not have intermediate values. To make the description closer to the human's perception we must augment binary movements with fuzzy variables to express the level of AU and of emotional state altogether. For example, "left eyebrow is a little raised" or "fully raised". These linguistic variables characterize different degrees of facial element manifestation. Now, the mimic state on the whole may be represented with a fuzzy vector of action units. To solve the task of BES recognition by this fuzzy vector the procedure of fuzzy inference is proposed. It is a well-known fuzzy classifier which may be realized as a hybrid neuro-fuzzy four-layer network of ANFIS type. The first layer contains the membership functions ("a little", "fully", "neutral", etc.). The second layer contains fuzzy neurons of logic conjunction. The third layer realizes the weightened linear combinations of the previous layer neuron outputs and the last layer is formed from the usual activation functions of sigmoid type. The classifier output is the basic emotional states such as "smile" or "sadness". The network may be taught with backward error propagation method, just like any usual neuro classifier.

Simpler then using linguistic evaluation variables is the interval evaluation methods which also may be proposed to express the degree of implementation of observed values to one of defined states. In this case, subjective evaluation should be omitted or statistical research must take place to define probability of belonging of observed state to evaluated emotional state.

An example of such research is represented in Fig. 1. Here, evaluated elements are mutual point positions of mouth, eyes, brows, nose and the face contour. The result of evaluation is a set of probabilities of one of the 7 BES (e.g. "happy with probability of 56 %").

Fig. 1. Emotional state recognition with the EmoDetect algorithm, (Neurobotics, 2013)

While using linguistic translations the statistical experiments become unnecessary, but subjective (or probabilistic) approach occurs in determination of functions describing fuzzy variable implementation (lips are compressed tight, mouth corners are a bit apart and so on). Nevertheless, undoubtful advantage of linguistic variable method is the possibility of more accurate analysis of facial expressions, which means evaluation of emotions, intermediate to BES.

The proposed method is bound to another approach, suggested in some works [1, 2] – the Emotional Cartesian Space (ECS). This approach is depicted in Fig. 2. In the ECS the x coordinate represents the valence and the y coordinate represents the arousal. Each expression e(v,a) is consequently associated with a point in the valence-arousal plane where the neutral expression e(0, 0) is placed in the origin. ECS allows describe an emotional state in more detail and should be used in further research.

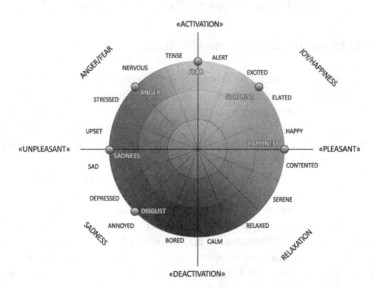

Fig. 2. Emotional Cartesian Space. BES are pointed on the edge of the circle.

Versus ECS the suggested approach in this article describes emotional state in terms of BES and can be represented as Fig. 3. ECS is used to solve the inverse task of evaluation of human facial expressions and was used in diagnostic research of several illnesses. The current analysis of the facial expression of operator is necessary to control the operator's state for safety of robotic or any other complicated control system.

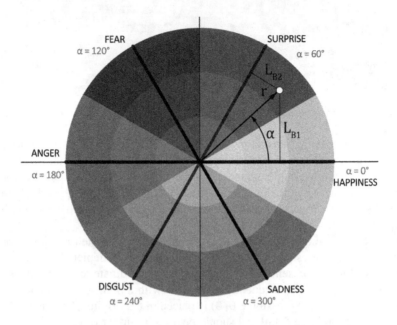

Fig. 3. Cartesian Space of Basic Emotional States

3   The Problem of Mimic Reproduction

Let's focus on the second group of problems – the reproduction (imitation) of mimic states for human-like robots. In this case the suggested scheme on Fig. 3 is more practical and efficient. This approach is suggested to be called as the Cartesian Space of Basic Emotional States (CSBES) [14]. In terms of interval method each sector corresponds to one BES and distinguishes its intensity ("force"). Usage of linguistic variables allows to determine emotion modality and intensity for each of BES.

The CSBES as well as the ECS uses interpolation of values between presets in each of the BES. A preset is an array of facial actuator positions that represent current emotional state. The radius represents the factor of emotion. The closer the point in CSBES to the central (neutral) point - the less aroused it is. The main difference between CSBES and ECS – the BES are positioned equally around the circle with a slight reorder. In the ECS the BES go as follows: Happiness, Surprise, Fear, Anger, Sadness, Disgust. In CSBES: Happiness, Surprise, Fear, Anger, Disgust, Sadness. This reordering has come from analysis of actuator arousal graphics, mainly the eyebrows and eyelids (Fig. 7). CSBES distributes the arousal evenly, without doubling the maximums of arousal functions. Thus, CSBES represents the facial actuator arousal model closer to the physical model of emotion reproduction.

There are several robotic emotional reproduction techniques: computer graphics, solid-state mechanics, multicolored facial backlight and anthropomorphic construction with skin imitation (most as a copy of infamous or existing people). Figure 4 shows examples of twin robot of H. Ishiguro, American robocopy of writer Philip K. Dick and Russian robot Alice Zelenogradova (Neurobotics Ltd.).

Fig. 4. Human-like robots: Geminoid (Japan), Philip K. Dick (USA), Alice (Russia)

The last robot is equipped with 19 actuators (Fig. 5) which represent the mimic apparatus for emotion synthesis. Actuators in this robot are servomotors and have following purposes:

- 3 servos implement the Gough-Stewart platform for the neck;
- 1 rotates the jaw to open the mouth;
- 3 ocular servos for mutual vertical and independent horizontal movement of cameras inside of each eye;
- Group of mimic servos deform silicone skin and take part in emotion synthesis:
  - 2 unpaired motors – "evil" and "puppy eyes" that accordingly move the point between the brows forward and upwards;
  - 10 paired motors – oral ("smile", "sad", "scepsis"), separate eyebrow and eyelid lift.

For example, fear is synthesized with mouth opening, « puppy eyes », left and right eyebrow lift and "scepsis" actuators.

Emotions are synthesized changing the actuator values (Fig. 6). The knowledge base of robot must contain settings that correspond to fuzzy emotional vector components, for each of basic emotional states. When receiving a command, expressed in fuzzy values (such as "strong anger" or "light surprise"), from higher level control system – the program select the corresponding emotional vector and linguistic transcription, that defines the intensity of emotion.

Coverage zones for different actuators do not match and the depiction of mimic state (in interval method) is asymmetric. The comparison of coverage zones in ECS and CSBES is shown on Fig. 7. In particular for the "evil" actuator, that brings the wrinkle between brows forward, in ECS we can observe 2 maximums and in CSBES – only one. Each maximum matches strong tension of corresponding actuator. There is a dead zone

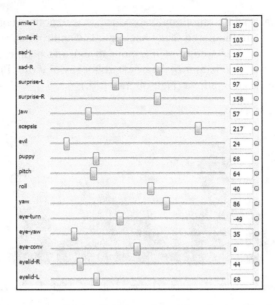

Fig. 5. Facial actuators of Alice Zelenogradova (Neurobotics, Russian Federation)

| **Neutral** | **Happiness** | **Surprise** | **Happy surprise** |
| CSBES: r = 0 | CSBES: a = 0°, r = 1 | CSBES: a = 60°, r = 1 | CSBES: a = 30°, r = 1 |

Fig. 6. Emotion synthesis of robot Alex (Neurobotics, Russia)

in ECS in lower right quadrant, the occurrence of which requires additional accuracy while defining the point of mimic state in other three quadrants. The CSBES (Fig. 4) has a different circumferential order of BES and lacks that dead zone disadvantage. This statement allows to synthesize emotional states more graphically and accurate with the same size of controlling elements on the screen.

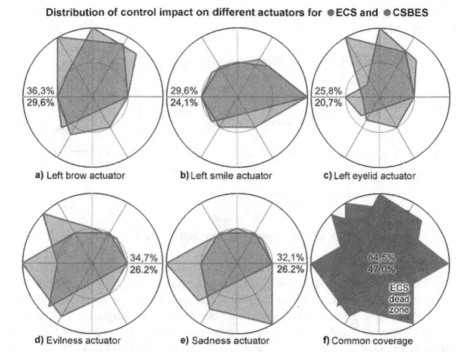

Distribution of control impact on different actuators for ●ECS and ●CSBES

a) Left brow actuator   b) Left smile actuator   c) Left eyelid actuator

d) Evilness actuator   e) Sadness actuator   f) Common coverage

Fig. 7. Difference of coverage zones (a-e) of ECS and CSBES of several actuators and common coverage zones (f) for every actuator. "A" means are of coverage of corresponding method. The outer bound corresponds to maximum, the center to minimum and the middle circle to the neutral positions of actuator.

For controlling the mimic apparatus of human-like robots there are several control methods, each of which has its advantages and disadvantages for different synthesis tasks. The independent BES control method has total control over modelled emotion, but its textual or encoded value is long and difficult to perceive. According a higher level method of ECS, the control signal is written short (2 coordinates), but requires some knowledge from the operator and accuracy in defining the 2D-point corresponding to the emotional state. Suggested method of CSBES is similar to ECS method, but is synthesized for graphical representation usability for an unprepared operator. This method allows to encode the control signal in formal mathematical and linguistic values and in more natural language record. CSBES method is compatible with several emotional state recognition algorithms and can be used for telemetric replication of mimic state between operator and a human-like robot.

4   Speech Interface and Dialogue

Speech communication is the main way for service robot control. The speech interface includes the recognition and linguistic modules (Fig. 8). The latter realizes semantic interpretation of information (speech understanding). The dialogue module controls the dialogue process.

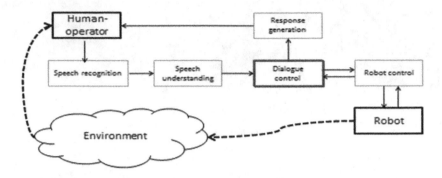

Fig. 8. The diagram of the speech interface.

Usually, speech recognition is based on the comparison of pronounced words with their patterns from a knowledge base. The most effective are the DTW (Dynamic Time Warping) and HMM (Hidden Markov Models). The speech base for service robot operators often may be individual. The dialogue may be presented as a sequence of messages. There are some typical dialogue scenarios. Messages are used: (a) to inform the operator about the automatically compiled plan of operation; (b) give the robot tasks by operator; (c) to correct the plan of operation or the condition of its fulfillment; (d) to inform about evaluation of environment by robot (e) to inform about results of operations; (f) to inform about evaluation by robot of its own condition, etc. The scenery itself may be presented as a sequence of frames.

In most cases, teaching by demonstration approach is used. [10]. Another way is the task-oriented learning [11, 15]. Some investigations in this field presents multimodal interfaces which combine speech and gestures [12]. Lately the authors demonstrated an alternative approach based on the modified Petri nets. The nodes of the net are the current states of the robotic system and the arcs are the events entering from other modules of the dialogue system. The kernel of the system is the dialogue manager which may address to the planning module to predict the next condition of the environment up to the desired condition [13].

Using the technology proposed above it is possible to connect the speech dialogue with the emotional content of the messages. The simplest way to solve the task is by replicating the mimic state of the operator's facial state during the speech message. More complicate task is to form the artificial mimic state of the robot "face" in connection with the sense of the information. Another block connecting speech understanding and emotion composition (modality and intensity of emotion) is introduced (Fig. 8). For example, negative emotion may correspond to the message of dangerous situation (for robot or for operator). The same is for message of operator's error during control or planning. Also for deficit of time for task solution. The positive emotion may be connected with successful operation fulfillment, optimal planning etc. Taking into consideration all the factors formalized in fuzzy terms it seems possible using the fuzzy procedure to determine the modality and the intensity of emotion. The interpretation of the corresponding emotion on the "face" of service robot make the mutual understanding in human-robot dialogue easier.

Together with the usual textual messages (from "speech-to-text") – the emotional messages can be used throughout the dialogue. The knowledge base of robot may contain reactions to emotional impacts (e.g. the human became sad or angry at some point of conversation).

5 Conclusion

The article proposed a new method of describing the changes of emotional states of human face in terms of actuator arousal. It enhances classic models of Emotional Cartesian Space and Action Units for the emotional reproduction with robotic "face".

Another approach for "naturalizing" robotic control is the speech interface. The speech interface is fuzzy, so as the emotional reproduction algorithm. They both work close together in natural speech interface.

The preliminary experiments showed that the combined speech – mimic interface make the human-robot dialogue much more productive and reliable. It most effective for service robotics as the operator uses the language close to natural. Usage of linguistic variables and fuzzy logic makes it possible to control robot by human almost without any special training.

The advances in appearance of robots (the problem of the "uncanny" valley soon will be solved), their interactive features and behavior lead to a new era of social robotics.

Now the next unsolved problem in human psychology is the contact with technical system almost in the same way as with a human, so the technical system becomes a partner. We suppose the intellectual interface is another step in the direction of new human-robot society.

References

1. Mazzei, D., Lazzeri, N., Hanson, D., De Rossi, D.: HEFES: a hybrid engine for facial expressions synthesis to like androids and avatars. In: 4th IEEE RAS & EMBS International Conference on Biomedical Robotics and Biomechatronics (BioRob), pp. 195–200 (2012)
2. Breazeal, C., Brooks, R.: Robot emotions: a functional perspective. In: Who Needs Emotions, 271–310 (2005)
3. Fominykh, I.: Techniques of solving creativity problems. In: Proceedings of 8th Artificial Intelligence National Conference "KII-2002" (2002)
4. Fominykh, I.: About an approach of emotion formalization. In: Proceedings of II International Scientific and Practical Seminar "Integrated Models and Soft Calculations in Artificial Intelligence" (2003)
5. Sholtz, J.: Theory and evaluation of human robot interactions. In: Proceedings of 36th Hawaii International Conference on System Sciences (HICSS'03) (2003)
6. MacDorman, K.F., Ishiguro, H.: The uncanny advantage of using androids in cognitive and social science research. Interact. Stud. 7(3), 297–337 (2006)

7. Ishiguro, H.: Design of Humanlikeness in HRI: from uncanny valley to minimal design. In: 8th ACM/IEEE International Conference on Human-Robot Interaction (HRI), pp. 433–434. IEEE (2013)
8. Ekman, P.: Facial expression and emotion. Am. Psychol. **48**(4), 384 (1993)
9. Ekman, P., Friesen, W.: Facial action coding system: A technique for the measurement of facial movement. Consulting Psychologists Press (2002)
10. Miura, J., Iwase, K., Shirai, Y.: Interactive teaching of a mobile robot. In: Proceedings of 2005 IEEE International Conference on Robotics and Automation, Barcelona, Spain, pp. 3389–3393 (2005)
11. Rogalla, O., Ehrenmann, M., Zöllner, R., Becher, R., Dillmann, R.: Using gesture and speech control for commanding a robot assistant. In: Proceedings of the 11th IEEE International Workshop on Robot and Human Interactive Communication (ROMAN), Berlin, pp. 454–459 (2002)
12. Kristensen, S., Horstmann, S., Klandt, J., Lohnert, F., Stopp, A.: Human-friendly interaction for learning and cooperation. In: Proceedings of the 2001 IEEE International Conference on Robotics and Automation, Seoul, Korea, pp. 2590–2595 (2001)
13. Zhonin, A.A.: An algorithm of learning of dialogue manager for a dialogue speech control system of robot. In: Proceedings of International Conference Integrated Models and Soft Computing in Artificial Intelligence, pp. 395–406 (2011)
14. Konyshev, D., Vorotnokov, S., Vybornov, N.: Control of mimic apparatus of service robots for emotion synthesis. Caspian J.: Manag. High Technol. **3**(27), 216–229 (2014)
15. Vorotnikov, S.A., Gorin, A.V., Konyshev, D.V.: Audio-visual sensor system for service robots. In: Proceeding of International Conference Extreme Robotics, pp. 246–252 (2010)

Multi-robot Exploration and Mapping Based on the Subdefinite Models

Valery Karpov[1,2(✉)], Alexander Migalev[1], Anton Moscowsky[1],
Maxim Rovbo[1], and Vitaly Vorobiev[1]

[1] National Research Centre "Kurchatov Institute", Moscow, Russia
karpov.ve@gmail.com, alexander.migalyov@mail.ru
moscowskyad@gmail.com, rovboma@gmail.com,
Vorobev_VV@nrcki.ru
[2] Moscow Institute of Physics and Technology, Moscow, Russia

Abstract. In this work we consider an environment exploration and mapping task for a group of heterogeneous mobile robots. We propose a range of methods tied together in a hierarchical two-level control system. The distinguished features of the proposed system are the use of subdefinite models for robot localization as well as an original mechanism for local interaction. We present both theoretical and experimental results. On the experimental side of the study we conduct both simulation experiments as well as a real robotic swarm system investigation.

Keywords: Collective robotics · Exploration · Mapping · Subdefinite models · Multi-robot systems

1 Introduction

Multi-robot systems have become a hot research topic due to the fact that many practical applications such as environment monitoring [1], technical inspection [2], search and surveillance [3] and several others can not be solved by a single robot. Many of these tasks can be formally expressed as various multi-robot exploration and mapping problems [4–6], which in turn can be decomposed into localization and mapping, path planning (obstacle avoidance) and multi-robot interaction (communication) problems.

Localization and mapping are traditionally viewed as a single coupled problem, e.g. SLAM problem, and various techniques and methods of solving this problem are known. Typically these methods rely on a Kalman filter [7], particle filter [8], graphs [9], etc. One should note that the vast majority of SLAM algorithms relies on accurate sensor measurements and meticulous odometry. This restricts the application of existing approaches to some well-defined classes of tasks and makes them unsuitable for other classes (for example, indoor SLAM methods are not good in outdoor navigation and vice versa). However, in general SLAM can perform well (or satisfactorily) if the SLAM pipeline is appropriately chosen in accordance with the features of specific tasks and robots.

In this paper we describe an approach of solving the multi-robot exploration and mapping problem which has the following distinctive features. Firstly, we rely on a

© Springer International Publishing Switzerland 2016
A. Ronzhin et al. (Eds.): ICR 2016, LNAI 9812, pp. 143–152, 2016.
DOI: 10.1007/978-3-319-43955-6_18

well-established SLAM methods but consider the output (a robot location and a local map) to be not error-free (which is a realistic assumption) and we augment this output with a topological map of the environment constructed using subdefinite models. Secondly, we rely only on the local robot interaction and use inter-robot communication to guide the path planning. As for the environment, we consider it to be partially known to the robots, e.g. we consider particular distinctive markers to be present (although neither their absolute, nor relative position is known a priori).

The paper is organized as follows. In Sect. 2 we present methods that are used to solve the multi-robot mapping and exploration task. In Sect. 3.1 we present the results of the modeled experiments. Section 3.2 is devoted to the investigation of the real robot case in indoor mapping scenarios.

2 Models and Algorithms

We consider the control system of each mobile robot involved in group exploration and mapping task to be decomposed into two hierarchical levels: strategic (high) and tactical (low). Navigation modules of the strategic level perform fuzzy localization of the robot using predefined markers and characteristic objects. A schematic map is constructed on this level via recognition methods based on subdefinite models calculations. Modules residing on the tactical layer deal with obstacle avoidance, trajectory following, etc. In this work we examine only the strategic level of the system, e.g. the methods that are used to localize a robot (by using subdefinite calculations and fuzzy spatial models), methods of map refining, mapping methods for a single robot being part of the group and, finally, map merging techniques and strategies.

Problem Statement. We consider a group of robots, each of which has the ability to determine its location and to mark obstacles on a map. Communication among the robots is strictly local — each one can talk only with its nearest neighbors. Therefore, the robots are equipped with transceivers with limited range. The goal of the robot group is to jointly construct a map of the environment.

As already mentioned, group or joint mapping tasks can be reduced to the problem of investigation of a map fragment by a single robot and global map construction from several such fragments obtained via communication between robots. The map fragment exchange task requires, as a rule, existence of a stable and broad communication channel. The communication system structure, in general, doesn't limit the robots' capability to transmit information to all accessible neighbors [10, 11]. However, the use of a fully connected topology is too bulky, despite its logical simplicity.

The global map construction can be reduced to the task of fragment integration, which consists of comparison of each part, similar areas detection and overlaying fragments correctly. Without the unified coordinates' binding it becomes a nontrivial task of searching for common subgraphs.

In the following section we describe the navigation task solution on the strategic level. Firstly, a mechanism that allows a robot to determine its location relying on a given schematic environmental map is described. After determining its approximate location, the robot should define the map more precisely. Then the movement trajectory planning mechanism is described, and finally the problem of group mapping and navigation is discussed.

2.1   Robot Localization

The robot localization system relies on data from a video camera (for landmark-based navigation) and from ultrasonic range sensors (for obstacle detection). Sub-definite procedures [12] are at the heart of the localization methods. These procedures are extensively used when the variables used to model some domain can not be determined precisely but only determined as probability interval estimates, which is our case. We represent robot's and markers' locations as sub-definite variables. It is also assumed within the approach under consideration that the interpolation functions that bind the variables' values with each other do exist. In addition, an iteration procedure is implemented that narrows the variables' domains at each step. Thus we work with a computational model which consequently narrows the area of indeterminacy of investigated variables' values, e.g. the robot's and markers' locations.

We use a well-known spatial model — a regular grid composed of square cells of identical size that represents the map. This representation is convenient and frequently used in such type of tasks, for example in [13]. As long as the localization system is based on landmarks (Fig. 1a), another approach is frequently encountered: along with registering cells on a map, objects and relations between them are also registered, for example [14, 15]. We use two-component color markers and marine signs of definite shapes as the landmarks (Fig. 1b). A recognition system was developed that is capable of estimating such attributes of the objects as color, form, size and orientation in space. An identification of objects was implemented that relies on a composition of various attributes.

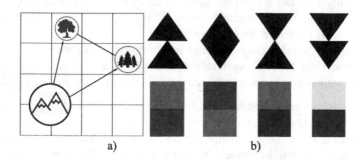

a)   b)

Fig. 1. (a) Combined map, objects and relations, (b) Used markers (Color figure online)

To identify the robot's location on a map we use a voting method. Suppose there exist several functions that determine possible locations of the robot based on several criteria: computer vision, sensor data, previous positions history, global navigation, etc. The robot's location in this case is identified by a distribution of candidate-cells r_{ij} on the map.

$$f^m(X_t) = R_{t^m} = \{r_{ij}, r_{ik}, \ldots\}, \tag{1}$$

where X_t is the input data at a moment t. Let the method 1 give the distribution R_1, method 2 — the distribution R_2, etc. Then the robot location is determined as follows:

$$R_t = k_1 R_t^1 \oplus k_2 R_t^2 \oplus \ldots \oplus k_M R_t^M, \tag{2}$$

where the operation $\oplus$ is a superposition of two distributions that in a simple case match a value to the amount of times the cell is presented in sets R^m, k_m — the weighting coefficient utilized for increasing influence of a criteria against the remainder.

$$R_t = \{c_{ij} r_{ij}, c_{ik} r_{ij}, \ldots\}, \tag{3}$$

where c_{ij} is the amount of times the corresponding cell is encountered with their weighting coefficients in distributions R_m. We will denote the value c in the following as the vote of this cell, and the function f_m — as a voting method. The system has three methods of voting implemented:

1. The cells from which the current scene can be seen are determined. In other words, for every cell the following condition is examined:

$$\left| dist(r(O_i), r_0) L(O_i) - dist(r(O_j), r_0) L(O_j) \right| < \delta, \forall O_i, O_j \in O(t), \tag{4}$$

where O_i is a current scene object $O(t)$, $r(O_i)$ is a map cell, containing an object O_i, r_0 — a verified cell, $dist(r_1, r_2)$ — the distance between the r_1 and r_2 cells, δ — an established error, $L(O_1)$ — the O_1 object's height on the image. The cells fitting this condition receive a voice.
2. An approximate distance from the robot to an object is determined based on the information about the approximate sizes of objects for every object. All cells positioned at this distance also receive a voice.
3. We assume that the robot can not change its position rapidly between two sequential observations. Thus, the cell, in which the robot was located in the previous moment of time, as well as neighboring cells, receives a voice in every step of the algorithm.

Consequently, we consider that the robot is located in the cell with the maximum amount of voices. A few cells could satisfy this criterion, so we choose only one using a mean value or using additional qualifying criteria. Let us describe the mechanism of voting in terms of subdefinite models (further SD-models). The robot position in this case is an H-value, i.e. some set, belonging to a whole range of definition. In this case the SD-expansion is the cells the map was divided into, and the range of definition accordingly is the whole set of map cells.

$$D_{NM} = \left\{ \begin{matrix} d_{11} & \ldots & d_{1M} \\ \ldots & \ldots & \ldots \\ d_{N1} & \ldots & d_{NM} \end{matrix} \right\}. \tag{5}$$

Let us use an SD-variable, determining a possible robot location as a cells set:

$$X = \{d_{ij} \dots d_{kp}\}. \tag{6}$$

The main idea of using SD-models is gradually decreasing the field of indeterminacy. The transformation of sets is performed using SD-operations.

$$f_i : X_1 \to X_2. \tag{7}$$

In case of the represented system, the SD-operations are voting methods, transferring all definition range to SD-values. Consequently, possessing a set of voting methods f_i and sequentially applying them to the initial set, we get a set of possible locations X_i for every method.

$$f_i : D \to X_i. \tag{8}$$

After that a superposition operator is applied, which leaves only the cells with the maximum amount of votes.

$$F : X_1 \dots X_i \to X_F. \tag{9}$$

The set represents possible robot locations. In the following, the averaging operation can be applied and one location among all others can be selected:

$$\bar{f} : X_F \to x. \tag{10}$$

2.2 Map Refinement

A map refinement procedure is triggered at each step of the navigation algorithm to update a partially known map in accordance with the new sensor measurements, e.g. measurements produced by the ultrasonic rangefinders. Each map cell is characterized by its weight w, which initially equals to zero and increases in case an obstacle is detected in that cell (so the cell is considered untraversable for the robot, Fig. 2).

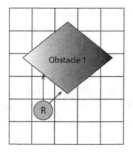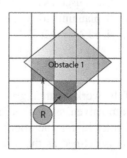

Fig. 2. Marking untraversable cells on a grid representation of the map.

As the location of the robot is subject to errors, the same is true for the obstacles. To handle this problem the following approach is adopted.

1. A degradation function F_D is introduced. This function decreases the weight of the cells over time in case these weights are below some predefined threshold. It means that cells with high probabilities of being blocked, e.g. the cells that are consequently reported as untraversable by the navigation system, will not lose their weight. At the same time, obstacles that were put on the grid map by mistake will tend to disappear. In the simplest case, a linear function can be used for cells weight degradation:

$$F_D(w) = w - k_D. \tag{11}$$

Here k_D is the degradation coefficient.

In case the current weight value is lower than the initial value (zero) the degradation procedure is not performed for this cell.

2. All possible untraversable cells (taking into account the robot's localization error) are marked so, but the weight for each cell is calculated by the following formula:

$$w_t = w_{t-1} + \frac{k_0}{N^2}, \tag{12}$$

where N is the number of possible robot's locations. Values of the coefficients k_0 and k_D are set in accordance with the following inequalities:

$$
\begin{aligned}
k_0 > \frac{k_D}{N^2}, \; for\; large\, N,\\
k_0 < \frac{k_D}{N^2}, \; for\; small\, N.
\end{aligned}
\tag{13}
$$

Thus, map refinement functions by receiving measurements of the rangefinders and a robot location on a grid map as the input and producing a marked grid as the output.

2.3 Individual Mapping and Map Merging

As written above, grids are used as environment models for each member of the group of mobile robots. Path planning for each robot is based on the following approach. Consider a grid area sized N × M:

$$F = (S_1, S_2, \ldots, S_{N \times M}), \tag{14}$$

Each element of the working area (grid cell) s_i is assigned the value of a potential φ_I which defines the attractiveness of that map area for a robot. Negative values are considered to be attractive for the robot, while positive — distractive. Movement direction in this case is defined as the direction of the whole field:

$$\bar{E} = -\sum_{i}^{N \times M} \overline{h_i}, \tag{15}$$

Here h_i is the derivative of the potential φ_i which in the simplest case can be calculated in the following way:

$$|h_i| = \frac{\varphi_i}{r_{i,robot}^k}, \tag{16}$$

where $r_{i,\,robot}$ is the distance between the robot and the considered (i-th) cell.

Unexplored regions of the working area are considered to be of high attractiveness for a robot initially and while it explores them they lose their attractiveness. During the exploration the robot always moves towards the area with the minimum sum of potentials. A merging procedure is triggered after all individual exploration tasks are finished. To merge maps, all robots must form a topology which we call a static swarm [16] and then start communicating with each other and exchanging maps and routes. To increase efficiency of the group exploration and mapping task we also suggest using local interaction.

Robots Interaction. If the robots are planning their paths to stay far from each other, unnecessary duplication of mapping effort (two or more robots mapping the same area) can be avoided. Thus, it can be beneficial to generate an "avoid me" signal so that the other robot that receives the signal knows that the area he is moving to has already been explored. At the same time, when the configuration of obstacles is not trivial, it can be beneficial to perform multiple examinations of such working area by different robots to obtain a consistent map model. In such cases, a communication command "approach me" should be utilized.

In accordance with the path planning strategy presented above "avoid me" and "approach me" signals can be implemented as changing the potential values of the corresponding grid cells, e.g. grid cell that are occupied by the robots generating the signals. For a robot's "avoid me" signal, the value of the cell's potential is increased (repulsing other robots) and vice versa. It should be emphasized, that these changes of the potential field are registered by each robot on its local map, depending on the cell that the signal was registered in by the robot. On the other hand, interaction between robots is implemented using explicit communication: robots broadcast messages to all their neighbours.

3   Experimental Evaluation

3.1   Model Experiments

Both in simulation and in real robot experiments the map was divided into a 30 by 50 grid of cells. Grids used in the experiments had 5 % to 30 % filling rate, meaning that 5 %–30 % of the cells were untraversable (occupied by the obstacles). Each experiment involved 16 robots. For each task, two communication profiles were used: with local communication while mapping and without such communication. As one can see from the Table 1, using the suggested local communication method significantly (an order of magnitude) increases the time efficiency of the mapping routine.

Table 1. Mapping time comparison

		Blockage percentage					
		5 %	10 %	15 %	20 %	25 %	30 %
Average mapping time (steps)	With communication	29,58	30,28	31,48	32,47	33,75	35,46
	Without communication	326,63	334,07	343,65	353,79	364,47	376,02

A step of simulation corresponds to the time a robot moves from a cell to an adjacent cell. One can note that the mapping time increases with the increase of the number of obstacles. It happens because a robot's path becomes more complicated when a larger number of obstacles is present. It could probably be possible to decrease the overall computation cost by using more advanced path planning techniques.

We would also like to note that, due to localization errors, the mapping method without local communication tends to detect more obstacles than actually exist. For example, for maps with 30 % blocked cells, the number of detected obstacles is one third times as much again than it should be. Using local communication while mapping positively influences this problem. In this case the number of detected obstacles almost equals the actual number.

3.2   Field Experiments

Field experiments were carried out using a DrRobot X80 Pro platform [17]. This robot is a two-wheeled mobile platform with a differential drive equipped with ultrasonic and infrared (IR) rangefinders as well as a Wi-Fi camera (Fig. 3a).

a)   b)   c)

Fig. 3.   (a) DrRobot X80Pro Platform used in the experiments, (b) Fragment of the polygon used in experiments, (c) Virtual model in Gazebo

The control system was implemented within the ROS framework: ROS Indigo [18] on a GNU/Linux Kubuntu 14.04 personal computer (installed on each robot). Control of a robot is based on automata techniques. All commands — both low-level (movement forward, backward etc.) and high-level (exploration, mapping etc.) are modeled as Mealy machines. CV system based on the OpenCV library is used for marker-based robot localization while ultrasonic range-finders data is used to detect obstacles and construct a local map. The global map is updated each time local maps are shared by the group.

The local communication was implemented on IR connection based on RC5 protocol. The aim of the local communication is the identification of the nearest robots that are considered neighbours and their relative position. Four transmitters and receivers are located on a robot facing different directions at a 90° angle. The robot transmitted its own unique ID number in four directions simultaneously through a standard RC5 command using four IR diodes.

Experiments were carried out on a polygon (Fig. 3) sized 10×20 meters with four robots involved. A virtual model of the polygon was created beforehand using Gazebo simulation software [19] in the ROS framework.

Twelve markers were placed in different locations on a polygon. Each marker is a cylinder with a marine navigation sign (unique for each marker) painted on the surface. The highest precision of localization was achieved when different markers were detected in a single image of the CV video stream. However due to the polygon large size, such situations occurred rarely. In this case, robots behaved in the following way. First robot performed a full rotation scanning all the markers and trying to localize itself by narrowing the ambiguity region. If this region still remained too large, then robot started to move in a random direction just following a trajectory created by the path planner and relying on encoders' measurements.

In case the markers are not unique, the localization error becomes much higher as the ambiguity region significantly enlarges. At this point, we would like to mention that a robot's location is composed of two components: a grid cell (x and y coordinates) and a heading angle (orientation). When the size of the grid cell was large (more than the diameter of a robot footprint), most of the errors occurred in the heading angle estimation. By latter we mean that the x and y coordinates' estimation error was much lower than the heading estimation error. This is due to the geometrical aspects of calculations as well as the low level of orientation discretization (8 values).

4 Conclusion

Conducted experiments showed the applicability of the proposed methods in exploration and mapping tasks. At the same time, the results of the simulated experiments only partially correspond to the real world experiments. This is due to the observation methodology of the experiments that allows giving only qualitative estimates.

Directions of the future work include but are not limited to the following. Firstly, the environment should become more complex by adding additional obstacles and active objects — intruders. In addition, a system of precise tracking of a robot's location on the polygon is needed. Existence of such a system will allow to obtain not only qualitative estimates but quantitative measurements, too.

Acknowledgements. The project was partially supported by RSF 16-11-00018, RFBR 15-01-07900 and RFBR 15-07-07483 grants.

References

1. Marino, A., Antonelli, G.: Experiments on sampling/patrolling with two autonomous underwater vehicles. Rob. Auton. Syst. **67**, 61–71 (2015)
2. Lopez, J., Perez, D., Paz, E., Santana, A.: WatchBot: a building maintenance and surveillance system based on autonomous robots. Rob. Auton. Syst. **61**, 1559–1571 (2013)
3. Portugal, D., Rocha, R.P.: Cooperative multi-robot patrol with Bayesian learning. Auton. Robots. **40**(5), 929–953 (2015)
4. Shvets, E.: Stochastic multi-agent patrolling using social potential. In: 29th European Conference on Modelling and Simulation, pp. 521–526 (2015)
5. Portugal, D., Rocha, R.P.: Multi-robot patrolling algorithms: examining performance and scalability. Adv. Robot. **27**, 325–336 (2013)
6. Elmaliach, Y., Agmon, N., Kaminka, G.A.: Multi-robot area patrol under frequency constraints. Ann. Math. Artif. Intell. **57**, 293–320 (2009)
7. Castellanos, J.A., Neira, J., Tardos, J.D.: Limits to the consistency of EKF-based SLAM. Robotics **45**, 1878–1881 (2004)
8. Montemerlo, M., Thrun, S., Koller, D., Wegbreit, B.: FastSLAM: a factored solution to the simultaneous localization and mapping problem. In: Proceeding of the 8th National Conference on Artificial Intelligence Conference Innovations Applications Artificial Intelligence, vol. 68, pp. 593–598 (2002)
9. De, A., Lee, J., Keller, N., Cowan, N.J.: Toward SLAM on graphs. In: Workshop on the Algorithmic Foundations of Robotics, pp. 1–16 (2008)
10. Burgard, W., Moors, M., Stachniss, C., Schneider, F.E.: Coordinated multi-robot exploration. IEEE Trans. Robot. **21**, 376–386 (2005)
11. León, A., Barea, R., Bergasa, L.M., López, E., Ocaña, M., Schleicher, D.: SLAM and map merging. J. Phys. Agents **3**, 13–23 (2009)
12. Narin'yani, A.S., Borde, S.B., Ivanov, D.A.: Sub-definite mathematics and novel scheduling technology programs. Artif. Intell. Eng. **11**, 5–14 (1997)
13. Elfes, A.: Using occupancy grids for mobile robot perception and navigation. Comput. (Long. Beach. Calif) **22**, 46–57 (1989)
14. Castle, R.O., Klein, G., Murray, D.W.: Combining monoSLAM with object recognition for scene augmentation using a wearable camera. Image Vis. Comput. **28**, 1548–1556 (2010)
15. Jensfelt, P., Ekvall, S., Kragic, D., Aarno, D.: Augmenting SLAM with object detection in a service robot framework. In: Proceedings of the IEEE International Workshop on Robot and Human Interactive Communication, pp. 741–746 (2006)
16. Karpov, V., Karpova, I.: Leader election algorithms for static swarms. Biol. Inspired Cogn. Archit. **12**, 54–64 (2015)
17. Dr Robot Inc: X80pro. http://www.drrobot.com/products_item.asp?itemNumber=X80pro
18. ROS Indigo Igloo. http://wiki.ros.org/indigo
19. Gazebo: Gazebo Simulator Official Site. http://gazebosim.org/

Positioning Method Basing on External Reference Points for Surgical Robots

Ekaterina Sinyavskaya[✉], Elena Shestova, Mikhail Medvedev,
and Evgenij Kosenko

Southern Federal University, Rostov-on-Don 347922, Russian Federation
esinyavskaya@sfedu.ru

Abstract. This paper proposes an automatic laparoscopic camera tracking for conducting optimal visualization of the required area such as operated field in the minimally invasive surgery. A robotic surgery system was designed and developed to perform the camera handling and tracking task during laparoscopic surgery. The method of positioning and automatic tracking for the surgical instruments during the laparoscopic operations was developed. The significant difference of the method is the usage of the markers and reference points that are placed at the visible area (outside the abdomen). This technique allows us to define the coordinates of the laparoscope and the surgical instruments in the operated field by the usage of the methods of vector algebra and geometric transformations without application of the image recognition. The algorithm of the laparoscope control and automatic tracking for the surgical instruments was offered. Also the conditions for the optimal visualization of the operated field were determined. The field of the required laparoscope position according to the surgical instruments was defined. The experimental research of the offered method was done and its justifiability was confirmed. The offered method of positioning and tracking is universal for the different types of the robotic holders with different number of DOFs.

Keywords: Robotic surgery system · Laparoscopic camera · The method of the positioning and tracking for the surgical instruments · Robotic camera holder for laparoscopy · Minimally invasive surgery · The reference points · Methods of vector algebra and geometric transformations · Position error

1 Introduction

The development in medical treatment is heading toward the minimization of surgical intervention and post-operative injury. It is possible by the usage of the minimally invasive surgery [1]. Visualization of the operated field is provided by the assistant actions based on voice commands of the surgeon. This approach has some disadvantages connected with reducing of depth perception of surgeon manipulation, increasing of hand tremor by the long operation, limited range of motion, increased fatigue and tactile limitations. Also there are some difficulties with control of laparoscopic camera position because of assistant tries to combine an overview of the treated area with the doctor's commands but it isn't always provide safe movement of the laparoscope in the

© Springer International Publishing Switzerland 2016
A. Ronzhin et al. (Eds.): ICR 2016, LNAI 9812, pp. 153–162, 2016.
DOI: 10.1007/978-3-319-43955-6_19

goal position. The more progressive usage of the robotic camera holders such as Aesop, EndoAssist, Naviot [2–5] can overcome some of these disadvantages because of the surgeon controls the camera via control interface by means of a pedal, joystick or voice control [6–9]. This kind of system does not exclude the influence of the human factor that can lead to errors by the surgeon as operator [10].

The development of the systems and methods based on the positioning and automatic tracking for the surgical instruments with the required accuracy allows us to take off task of camera control from assistant and surgeon and therefore decrease the influence of the human factor, reduce the number of medical staff and extend the functionality of the surgeon during the operation.

2 Description of the Robot-Assisted Surgical System

The process of the laparoscopic surgery by the usage of the robotic laparoscopic camera holder is represented as a system. In Fig. 1 the flow chart of the robot-assisted surgical system with automatic positioning and instruments tracking is shown [10]. Initially the positioning process is described only for one surgical instrument, as the placement of the second (auxiliary) surgical instrument is defined similarly to the first, while the changing of the laparoscope position in the operated field is fulfilled relatively to leading tool [4–8, 11].

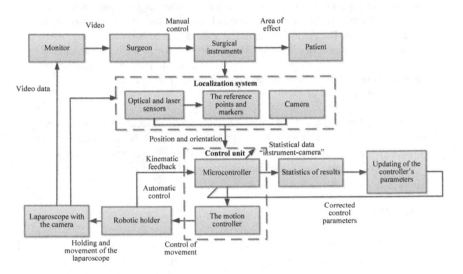

Fig. 1. Flow chart of the robotic surgical system

The surgeon performs operation basing on the video image of the operated field by the usage of the surgical instruments that are introduced by the trocars. The current coordinates of the laparoscope and instruments are defined by the usage of the optical localization system and external video camera in the operating room. This kind of localization system uses the external reference points and markers that are placed on the visible parts of trocars. The position and orientation of the laparoscope and instrument are determined by the angle of the reflected rays from the reference points.

The obtained coordinates come in the control unit where the microcontroller generates the control commands for the changing of the position and orientation of the robotic holder or focusing of the laparoscopic camera in order to get the optimal visualization of the operated field. The control unit transmits the generated control signals to the motion controller. The motion controller sends the commands to the robot actuators for changing of placement of the robot joints according to the required trajectories of the joints and goal position of the laparoscope.

All decisions that relate to certain input conditions (placement of the surgical instruments) and output conclusions (characterize the movements of the laparoscope) are kept into the block "statistics of results". The information about all results is used for comparative analysis of work and following adaptation of control unit parameters. The robotic holder moves the laparoscope depending on the field of the surgical operation and current position of the instruments. The kinematic connection between robot links during its movement is necessary to determine the boundaries or safe field of function.

Further image of the operated field from the laparoscopic camera is transmitted to monitor via communication links that allows surgeon to track for his/her actions and continue to perform the operation in real time.

The offered system of automatic tracking for surgical instruments can be used for the different commercial robotic holders of laparoscope and robot-assisted surgical systems, such as Aesop, Zeus, SoloAssist, da Vinci. Existed robotic holders have different DOF, sizes of the links and types of joints. But the usage of the offered automatic control system of the laparoscopic camera doesn't depend on the robot configuration [12–15].

3 Method of Positioning and Automatic Tracking for the Surgical Instruments Basing on the Markers and Reference Points

The goal of the offered method is connected with determination of the placement of surgical instrument and laparoscope in the operated field basing on the visible parts outside tools [16–18]. If the measurements of the instrument and laparoscope are known in advance it can be determined their position and orientation in the operated field by the application of the mathematical methods without analysis and image recognition. In Fig. 2 the reference points and DOF of the surgical instrument and laparoscope are shown for the minimally invasive surgery.

In Fig. 2 there are following accepted designations: O_1 – laparoscope incision point; O_2 – surgical instrument incision point; A – end effector of the instrument; B – end effector of the laparoscope; C – tip of the surgical instrument; D – tip of the laparoscope; ρ – length of the visible part of the laparoscope; Θ – angle between the laparoscope shaft and the vertical axis; φ – rotation angle.

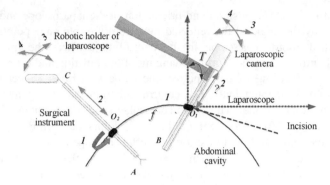

Fig. 2. Surgical instrument and laparoscope during minimally invasive surgery

Both angles are described according to the pivot point (incision point). The point O_1 is determined as origin.

The position of the surgical instrument and laparoscope is determined basing on the triangulation method by the application of optical system such as Polaris [19] that uses the LED sensors on the laparoscope and surgical instrument for definition of the points coordinates C_i, D_i, O_1, O_2 [20].

The difference of the offered approach is that the coordinates of the reference points of the laparoscope and instrument are determined in the visible area basing on the receiver fixed in advance in the operated room [17, 18, 20]. Basing on the common view of the operated room the coordinates of the reference points C, O_2, D, O_1 are found sequentially from sources (L_1, L_2, L_3) and receivers (S_1, S_2, S_3) (Fig. 3).

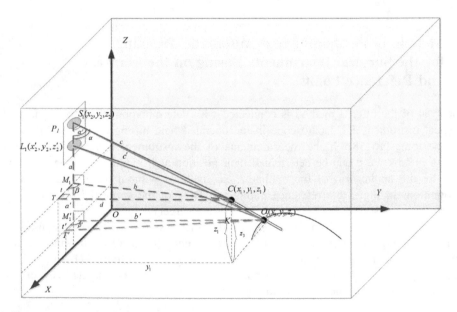

Fig. 3. The process of the transmission signal from the reference point C and receiver S_1

The coordinates of the A and B are calculated by the usage of the spherical coordinate system after determination of the coordinates of the points C_i, D_i, O_1, O_2 and orientation of the corresponding vectors [21]:

$$x = \rho sin\Theta cos\varphi, y = \rho sin\Theta sin\varphi, z = \rho cos\Theta, \rho \geq 0, -\pi \leq \varphi \leq \pi, 0 \leq \Theta \leq \pi. \quad (1)$$

The task of the automatic tracking for the surgical instruments and providing of optimal visualization of the operated field is reduced to determination of the coordinates (x_1, y_1, z_1) of the point A and basing on this point the required position and orientation of the laparoscope is defined (x_2, y_2, z_2) of the point B (Fig. 4). The optimal view of the operated area is determined by the position of the instrument end effector that should be in the center of video laparoscope image transmitted to the monitor.

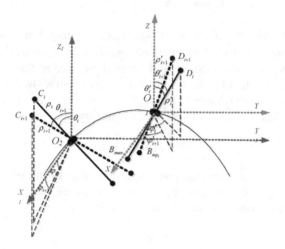

Fig. 4. Schematic representation of the tracking task

There are two conditions for reaching of optimal view of the operated area:

- the lines which are located the vectors $\overrightarrow{CA}$ and $\overrightarrow{DB}$ must be intersecting, that is $\left(\overrightarrow{CA} \cdot \overrightarrow{DB}\right) = 0$;
- the vectors $\overrightarrow{CA}$ и $\overrightarrow{DB}$ must be noncollinear, the point A_i – the center of the lines intersecting which are located the vectors $\overrightarrow{CA}$ и $\overrightarrow{DB}$.

The offered algorithm of tracking for the surgical instruments consists in the following steps.

Step 1. The coordinates (x_5, y_5, z_5) of the point O_1 (laparoscope incision point) and (x_6, y_6, z_6) of the point O2 (surgical instrument incision point) are determined.

Step 2. The coordinates (x_1, y_1, z_1) of the current position of the end effector of the instrument A_i, $i = \overline{1; N}$ at time point $t = i$ are determined.

Step 3. The equation of line is constructed by the coordinates of the points A_i and O_2. The surgical instrument lies on this line. The canonical equation of the line has the following view [21, 22]:

$$\frac{x - x_6}{x_1 - x_6} = \frac{y - y_6}{y_1 - y_6} = \frac{z - z_6}{z_1 - z_6}. \tag{2}$$

Step 4. The equation of line is constructed, on this line the laparoscope lies at time point $t = i$. The equations of lines for the surgical instrument and laparoscope intersect at point A_i. The equation of line is constructed by the coordinates of the points A_i and O_1. The canonical equation of the line on which the laparoscope lies has the view:

$$\frac{x - x_5}{x_1 - x_5} = \frac{y - y_5}{y_1 - y_5} = \frac{z - z_5}{z_1 - z_5}. \tag{3}$$

Step 5. The coordinates of the point $B_{req_i}\left(x_{B_{reqi}}, y_{B_{reqi}}, z_{B_{reqi}}\right)$ at time point $t = i$ are determined. The optimal view of the operated field is determined by the position of the end effector of the instrument that should be in the center of the video laparoscope image. Therefore restriction on the length of the maximum allowable area of the end effector of the laparoscope relative to the position of the end effector of the instrument depends on the safety settings and visualization quality:

$$k_{min} \le |B_{max}B_{min}| \le k_{max}, \tag{4}$$

where k_{min} and k_{max} – the minimum and maximum value of the length $B_{max}B_{min}$.

Condition of the restriction on the allowable distance from the instrument end effector to the laparoscope end effector has the following view:

$$r_{min} \le |r| \le r_{max}, \tag{5}$$

where r_{min} and r_{max} – the minimum and maximum allowable distance from the end effector of the instrument to the end effector of the laparoscope.

Step 6. Scope of the operated field is determined also by the viewing angle of the camera. The laparoscopic camera scope depends on the laparoscope configuration but it can be represented in common view as shown in Fig. 5 [23].

Angle θ between the vectors V_{CA} and V_{AB} is determined as [22]:

$$\theta = \cos^{-1}\left(\frac{\vec{V}_{CA} \cdot \vec{V}_{AB}}{|V_{CA}| \cdot |V_{AB}|}\right), \tag{6}$$

If the value of the angle θ is less than θ_{req} than it means that the instrument locates in the scope (video laparoscope image). If the value of the angle θ is more θ_{req} than it means that the instrument locates out of the video laparoscope image.

Step 7. The position error is calculated after fulfillment of tracking process and it is represented as deviation between the current and required position of the laparoscope:

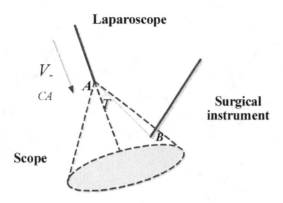

Fig. 5. Laparoscopic camera scope

$$e_i = \sqrt{(x_{B_{req}} - x_{B_{cur}})^2 + (y_{B_{req}} - y_{Bcur})^2 + (z_{B_{req}} - z_{B_{cur}})^2} \leq 1 \, \text{mm}, \qquad (7)$$

where $(x_{B_{req}}, y_{B_{req}}, z_{B_{req}})$ – the coordinates of the required position of the laparoscope; $(x_{B_{cur}}, y_{B_{cur}}, z_{B_{cur}})$ – the coordinates of the current position of the laparoscope.

Step 8. Transfer is fulfilled t to step 2 at the next moment $t = t + 1$.

Step 9. The algorithm continues until the operation is complete or end of work the operation in the automatic mode.

Step 10. The mean square deviation for every link is calculated after end of work of the algorithm:

$$\sigma = \sqrt{\frac{1}{N} \sum_{i=1}^{N} (e_i - \bar{e})}, \qquad (8)$$

where σ — mean square deviation of the position error; e_i — the current value of the error; $\bar{e}$ — the mean value of the error.

4 Results of the Experiments

The 20 tests of the different types of trajectories were fulfilled in the programming language Matlab R2014b. In the Fig. 6 the path following graph of current and required trajectories of the laparoscope for the first test is shown. The current trajectory of the laparoscope movement approximates to the required characteristic. The maximum value of the position error was 1 mm and the minimum value was 0.1 mm.

The values of the position error consist only in deviation of the method, configuration parameters of the robotic holder and hardware in this research are not taken into account. The mean square deviation of the position error by the 20 tests was 0.3781 mm. Therefore the maximum value of the position error does not exceed the minimum threshold by the parameter of the positioning accuracy [4, 5, 7, 8, 11, 17].

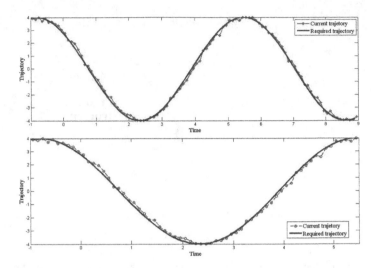

Fig. 6. The path following graph of the current and required trajectories of the laparoscope for the first test

The obtained value of the mean square deviation with taking into account hardware with a high probability allows providing the given value of positioning accuracy. The values of the position error and mean square deviation do not exceed the threshold value (7). Basing on the source analysis the threshold of position error is 5 mm [11, 17].

5   Conclusion

The method of the positioning and automatic tracking for the surgical instruments during the minimally invasive surgery was developed. The difference of this method is the usage of the markers and reference points that are placed on the visible area. This feature allows to determine the accurate position and orientation of the instrument and laparoscope in real time, whereas the most approaches based on the intraabdominal analysis of the operated field where bad visibility and heterogeneity of the environment decrease accuracy of the obtained data.

The algorithm of the automatic tracking for the surgical instruments with the usage of the vector algebra and geometric transformations was developed. Experimental research with the different types of the trajectories was done. The received values of the error are not exceed the threshold value (7) therefore the positioning requirement for the surgical instrument is satisfied. The offered method can be used in the other fields of robotics where necessary to provide autonomous camera control and high accurate process of positioning.

Acknowledgement. This paper was made with support of the Russian Science Foundation Grant 14-19-01533 at the Southern Federal University, Russia.

References

1. Fan, G., Zhou, Z., Zhang, H., Gu, X., Gu, G., Guan, X., Fan, Y., He, S.: Global scientific production of robotic surgery in medicine: a 20-year survey of research activities. Int. J. Surg. **30**, 126–131 (2016)
2. Robotics. http://robotics.com.ua/shows/series_robots_and_humans/3345-your_health_health_robotics_today
3. Medical robots. http://medrobot.ru/744-pechalnye-oshibki-medicinskih-robotov.html
4. Sim, H.G., Yip, S.K.H., Cheg, C.W.S.: Equipment and technology in surgical robotics. World J. Urol. **24**(2), 128–135 (2006)
5. Kommu, S., Rane, A., Eddy, B., Rimington, P., Anderson, C.: Initial experience with the endoassist (r) camera holding robot in laparoscopic urological surgery. Eur. Urol. Suppl. **6**(2), 186 (2007)
6. Baik, S.H.: Robot Surgery. InTech, Vienna (2010)
7. Taylor, R.H., Kazanzides, P.: Medical robotics and computer-integrated interventional medicine, p. 24. Elsevier (2008)
8. Dumpert, J.J.: Towards supervised autonomous task completion using an in vivo surgical robot: dissertation. University of Nebraska, Lincoln (2009)
9. Dalela, D., Ahlawat, R., Sood, A., Jeong, W., Bhandari, M., Menon, M.: The growth of computer-assisted (robotic) surgery in urology 2000–2014: the role of Asian surgeons. Asian J. Urol. **2**, 1–10 (2015)
10. The potential of Russian innovations at the market of automation and robotics systems. Expert-analytical report. RVC OJSC, Moscow (2014). https://www.rusventure.ru/ru/programm/analytics/docs/Otchet_robot-FINAL%20291014.pdf
11. Jaspers, J.: Simple tools for surgeons. Design and evaluation of mechanical alternatives for robotic instruments for minimally invasive. Dissertation, Amsterdam (2006)
12. Fahimi, F.: Autonomous Robots Modeling, Path Planning, and Control. Springer, Heidelberg (2009)
13. Zenkevich, S., Yushchenko, A.: Fundamentals of Control Robotic Manipulator. MSTU Bauman, Moscow (2004)
14. Moskvichev, A., Quartalov, A., Ustinov, B.: Gripping devices of industrial robots and manipulators. A tutorial. Forum (2015)
15. Pshikhopov, V.K., Medvedev, M.Y., Kostjukov, V.A., Gaiduk, A.R., Fedorenko, R.V., Gurenko, B.V., Kruhmalev, V. A., Medvedeva, T.N.: Design of robots and robotic system. A tutorial. SFU, Taganrog (2014). http://ntb.tgn.sfedu.ru/UML/UML_5248.pdf
16. Bogdanova, Y.V., Guskov, A.M.: Numerical simulation of the positioning task of the surgical tool of robot manipulator while moving along the predetermined path. Electron. Sci.-Tech. J. «Sci. Educ.» (5), 181–210 (2013)
17. Agustinos, A., Wolf, R., Long, J.A., Cinquin, P., Voros, S.: Visual Servoing of a robotic endoscope holder based on surgical instrument tracking. In: 5th IEEE RAS and EMBS International Conference on Biomedical Robotics and Biomechatronics (BioRob), Sao Paulo, Brazil, pp. 13–18 (2014)
18. Berkelman, P., Cinquin, P., Boidard, E., Troccaz, J., Lretoublon, C., Ayoubi, J.-M.: Design, control and testing of a novel compact laparoscopic endoscope manipulator. Proc. Inst. Mech. Eng.: J. Syst. Control Eng. **217**(Part I), 329–341 (2003)
19. Official website of Polaris (Electronic resource). http://www.ndigital.com/

20. Finaev, V.I., Sinyavskaya, E.D., Shestova, E.A., Kosenko, E.Yu.: Design of the positioning method of the robotic holder of laparoscope basing on computational geometry. Izvestia SFedU. Technical Science, **2**, Taganrog (2016). http://izv-tn.tti.sfedu.ru/wp-content/uploads/2016/2/6.pdf
21. Beklemishev, D.V.: Course of Analytical Geometry and Linear Algebra, 10th edn. Fizmatlit, Moscow (2005)
22. Ilyin, V.A., Kim, G.: Linear Algebra and Analytic Geometry. Publishing House of the MSU, Moscow (1998)
23. Intelligent autonomous camera control for robotics with medical, military, and space applications, no. US 20130331644 A1 (2010)

Preprocessing Data for Facial Gestures Classifier on the Basis of the Neural Network Analysis of Biopotentials Muscle Signals

Raisa Budko[✉], Irina Starchenko, and Artem Budko

Southern Federal University, Taganrog, Russian Federation
raisa-budko@ya.ru, {star,abudko}@sfedu.ru

Abstract. The recognition of facial gestures using biopotentials muscle signals has been proposed for human machine interface. Real-time myoelectric control requires a high level of accuracy and computational load, so a compromise between these two main factors should be considered. The most informative electromyogram features, required number of channels and the most suitable architecture of the neural network were identified in this study. In this paper, a results of preprocessing data were proposed to use in facial gestures classifier. The effectiveness of different sets input data combinations was also explored to introduce the most discriminating.

Keywords: Electromyography · Gesture recognition · Neural networks · Human machine interface · Signal preprocessing

1 Introduction

Facial gesture recognition is important for touch free control of different devices, such as wheelchairs, artificial limbs, and so forth. Despite the fact that there are a number of publications on this problem [1–4,7], some areas still have a wide space for research. For example, an equivalent of this method is the use of camera. This method has several drawbacks and limitations. Firstly, changes in the light conditions makes impossible recognition of facial gestures (except for infrared (IR) camera, but it has a high cost). The proposed method is independent of lighting conditions. The second, using camera in gesture recognition requires large computational resources. Under the same technical resources, the proposed method has the best performance. That's why the purpose of this research is the analysis of one particular area the assessment of capability for facial features gesture recognition based on the study of on the facial EMG signal characteristics.

2 Methodology

2.1 Data Acquisition

The experimental part of the study is implemented by using the following equipment and software: "Synapsis" system (developed by SMC "Neyroteh",

© Springer International Publishing Switzerland 2016
A. Ronzhin et al. (Eds.): ICR 2016, LNAI 9812, pp. 163–171, 2016.
DOI: 10.1007/978-3-319-43955-6_20

registration certificate FSR 2010/07176, March 29, 2010; software for electroneuromyographic system "Synapsis" (developed by SMC Neyroteh). Post-processing and signal detection are performed by using MATLAB (developed by The MathWorks).

Surface one-use electrodes (1 cm in diameter located at 2 cm distance) are used for the acquisition of electromyographic signal. Recording configuration is bipolar, to reduce the influence of the any noise component. Electrodes are placed on the frontalis muscles, dexter temporalis and sinister temporalis muscles [7].

Ten mentally and physically healthy volunteers (5 males and 5 females aged 19–26 years) participated in the study. Mimic gestures considered in this study are the following: (1) jaw contraction; (2) "ear-to-ear" smile; (3) raising eyebrows; (4) frowning eyebrows. The skin was cleaned using alcohol wipes to remove grease and traces of sweat. Electrodes are placed on muscle belly in order to achieve the higher amplitude signal [6,7].

The participants made each gesture 5 times in 2 s (active signal) with a 5 s pause between the muscle contraction to eliminate muscle fatigue effect. The power spectrum was calculated for the evaluation of the signals for each component of the noise signal for each participant.

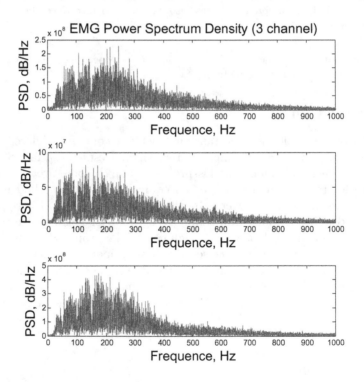

Fig. 1. The power spectrum for a sample of one of the participants

EMG analysis is implemented to extract the features that allow to classify gestures in this study. So it was necessary to eliminate the effect of any noise on the curve shape.

The graph of the power spectrum for one of the participants on the sample "jaw contraction" is shown in Fig. 1. Slices can be seen at frequencies of 50, 100, 150, 200, 250 Hz. It occurred because of the usage of notch filter. As for the rest the power spectrum for facial EMG is trivial. EMG bandwidth is placed in the range from 0.1 Hz to 400 Hz, which corresponds with a range of facial EMG. Then the pre-processing of the signal has been carried out for the removing of peak-burst and data smoothing.

2.2   Active Signal Extraction

It is necessary to extract the active signal part, because only active signal is useful for the detection and determination of various facial gestures. Despite the fact that the sample was recorded in the "5 s rest - 2 s force" mode, there are areas of active signal that trespass the 2 s interval, (Fig. 2).

Active signal interval may either achieve higher or lower step of the defined interval or be dislocated in terms of a determined time period. This is due the inaccurate performance motions member. Taking this into account, 7 s of active signal are used for further processing. Signals are recorded on the three channels simultaneously, in result there is a three-dimensional data set (with 3×7) for each gesture.

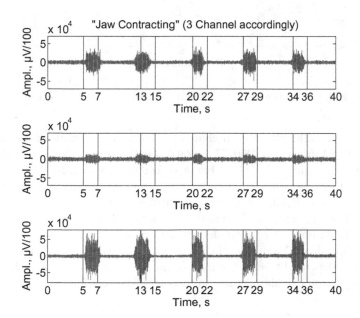

Fig. 2. Screen "Jaw Contracting" in MATLAB. Dotted lines are plotted on the intended signal which should be located

2.3   The Input Feature Vector Creation

It is more rational to consider not the very indications of the time series, but its most essential features. That's why, the filtered signals were segmented at the preparatory feature selection stage with non-overlapping windows of 25 ms length [4,8].

Figure 3 presents the results of the segmentation and feature extraction for one of the study participants.

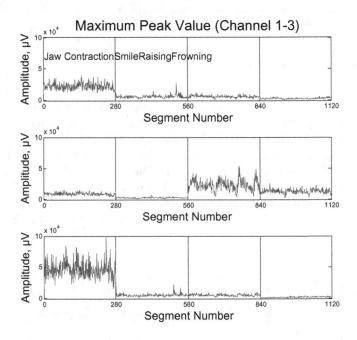

Fig. 3. Creating a feature vector according to MAX feature

Since the signal length of each channel is 7000 ms, in result of segmentation we obtained 280 intervals (7000/25 = 280). Feature extraction is an important step in the treatment of EMG. It has a direct impact on final performance of the system. Reliable signs should allocate the most important characteristics of the facial EMG signal. They also should have a low computational cost for use in real-time applications. Previous studies have been investigated a number of different functions with different complexity and efficiency for EMG signals processing [5,6]. Peak Value was the most discriminative feature for the number of studies [8]. It is used to find the maximum absolute peak value of EMGs. For further analysis we calculated maximum peak of EMG, defined as:

$$x_k = max|x_i|, \tag{1}$$

where k - segment number, x_i - the instantaneous value of EMG signal segment [5].

As a result, three-dimensional function-vector was calculated for each participant of the study, which contained 1120 values (for four facial gestures).

2.4 Minimum Redundancy - Maximum Relevance Criterion

The authors suggest MRMR criterion to evaluate the efficiency of each channel [7]. Algorithm MFMR (minimum redundancy - maximum relevance) uses a heuristic criterion to establishing a balance between the most relevant (bond strength characteristics of the response), and the minimum excess (connections between pairs of features). This is a greedy algorithm (based on the selection of a particular one pass), which focuses only on pairwise redundancy and ignores conjugate (compound logical connections features to predict the answer).

The reducing of the computational complexity is the main motivation for reducing the number of channels. In addition to the channels, which have a low classification ability is very likely situation of two "good" canal (with almost equal classifying abilities), which have high correlation with each other. The second reason to reduce the number of information is the increase of the classifier's generality. Minimum redundancy criterion is often used for it.

The maximum relevance is the choice of signs in pattern recognition, which aims to identify the subset of data that are relevant and best describe the statistical properties of the classified variable. MRMR method was proposed nearly a decade ago [7].

Average values of all Mutual Information values between individual functions fi and class C determine the relevance of feature set A and class C as follows:

$$D(A, C) = \frac{1}{|A|} \sum_{f_f \in A} MI\left(f_i, C\right), \tag{2}$$

where MI mutual information, is calculated as:

$$I(A, B) = \sum_{b \in B} \sum_{a \in A} p\left(a, b\right) \log \left(\frac{p\left(a, b\right)}{p\left(a\right) p\left(b\right)}\right), \tag{3}$$

where $p(a, b)$ is the joint probability distribution function of A and B, $p(a)$ $p(b)$ is the marginal probability density functions A and B respectively.

It is seen that $I(A; B) = 0$, where A and B are statistically independent and $p(a, b) = p(a)p(b)$. This means that the combination of several individual attributes can achieve high recognition accuracy if they provide additional information.

The redundancy of all features in set A is calculated as:

$$R(A) = \frac{1}{|A|^2} \sum_{f_f f_j \in A} MI\left(f_i, f_j\right). \tag{4}$$

Then MRMR can be achieved by finding a maximum value of the difference of these parameters [7]:

$$MRMR = \max\left[D\left(A, C\right) - R\left(A\right)\right]. \tag{5}$$

2.5 Classification

In order to recognize facial gestures, the extracted features are to be classified in distinct classes. The classifier should not depend on factors that have a significant impact on the EMG patterns over time, such as the displacement of the electrode, movement artefacts of EMG signals, skin excretions. Also new classifier should classify samples with sufficient accuracy in real time with low cost calculations. This is an essential condition for the human machine interface systems.

According to previous studies we know about successful implementations of classifiers based on neural networks for classification myoelectric features. In this study we suggest the use of neural network based on radial basis functions for the classification of facial EMG. This method was proposed by S. Jaiyen and its robustness was tested and proven by multiple data sets [8].

This routine training is very fast compared to traditional neural networks, such as the direct distribution network, and a relatively small amount of memory required for it [5–7]

Radial basis function networks have a number of advantages compared to multi-layer feedforward networks. Firstly, they model arbitrary non-linear function with a single intermediate layer. This saves the developer from need to decide for the number of layers. Second, the parameters of the linear combination of the output layer are optimized by well-known linear optimization methods. These methods are fast and there are no problems with local minimum, which interfere with the learning algorithm back propagation. So RBF network is trained much faster than a back-propagation algorithm.

The neural network was created with an RBF algorithm after the development of input feature vector. The network architecture is the following: the number of hidden neurons - 21, learning algorithm - radial basis function, error function - entropy, activation function - Gaussian. Output activation: Softmax function. Softmax together with a cross-entropy error function allows to modify the network for probability estimate of class affiliation. The network was trained for the half the original data [2].

3 Results and Discussion

3.1 Classification and Recognition Accuracy

Classification and recognition accuracy are shown in Table 1. This result was obtained using the three data channels to all participants as the input feature vector, and the average results indicating the error and the standard deviation.

Average maximum recognition accuracy was 93.4 % on the testing stage. The best and the worst results are listed in the column "Maximum (Test)" and "Minimum (Test)" for each of the study participants throughout the testing. Research subjects 1, 2, 3, 4, 5, 6, 7 and 10 reached the maximum recognition performance (more than 90 %).

Table 1. Classification and recognition accuracy for each participant and for three input channels

Feat. \ Pat.		1	2	3	4	5	6	7	8	9	10	Av.±Std	Err., %
MAX	Train	98,5	98,0	97,0	92,3	98,0	93,7	98,0	81,2	86,0	93,3	93,6±5,8	6,4
	Test	98,5	98,0	95,0	93,5	97,0	94,5	95,0	86,5	84,0	92,0	93,4±4,8	6,6

3.2   The Input Feature Vector Dimension Analysis

This section describes a ranking by criterion MRMR channels. It also evaluates the impact of combinations of channels on the performance of the neural network based on different input signal length, which used for network training. MRMI selects channels based on the value of the mutual information measure that is based on the degree of correlation between the functions themselves.

As a result channel 2 showed the highest rank of importance (its electromyogram was recorded from frontalis muscle). Channel 3 showed the lowest rank (its electromyogram was recorded from sinister temporalis muscle) after channel 1 (from dexter temporalis muscle).

Analysis was made on the construction of a network with a different dimension of the input feature vector. Two kinds of combinations have been assessed, with using full-time active signal, and using only first second of every gesture. Table 2 presents the studied combination for the channel number signal time:

Table 2. Sets for the time signal and channel number combinations and its training results

N	The channel number (the vector dimension)	Time, s	Accuracy, %	Training time, s
N1	3 channel (channel_1; channel_2; channel_3)	7	93,4	0,25
N2	2 channel (channel_1; channel_2)	7	94,5	0,18
N3	1 channel (channel_2)	7	85,6	0,31
N4	3 channel (channel_1; channel_2; channel_3)	1	97,2	0,16
N5	2 channel (channel_1; channel_2)	1	97,1	0,14
N6	1 channel (channel_2)	1	73,3	0,35

Figure 4 shows the performance and the training time of achieved networks, averaged for all of the participants. Networks were trained for three, two and one channel respectively with the signal time of 1 and 7 s (according to Table 2).

Recognition accuracy was about the same for sets N1 and N2 (93.4 % and 94.2 % respectively). Also, it was for sets N4 and N5 (96,2 % and 95,1 % respectively). It means the same information that is contained in the temporal muscle channels. Recognition accuracy was low for both 7, and for 1 s input channel

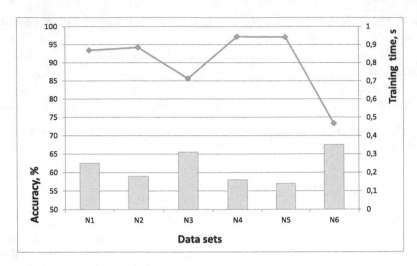

Fig. 4. Influence of channel numbers and input feature vector time on the recognition accuracy and training time

while using only one channel, which showed the highest rank of significance (N3 and N6, 85.6 % and 73.3 % respectively). In addition, the speed of learning has grown strongly for it. So, despite the fact that the second channel showed the highest rank value, it cannot provide enough information for classification without other channel.

The next explanation has been offered to explain these results. The each separate channel carries information from other muscles but a combination of the three channels provides less discriminative feature sets. Also it caused more data overlapping between the output classes which reduced the classification accuracy. It can be concluded that the use of two channels optimally in terms of time training and performance. Furthermore, using a one second of the input signal provides a gain in training time.

Compared to another study on this problem [8], where the similar gestures were examined, we noted that the classifier, which was used there, has lower speed of training and higher dimension vector of input features.

4   Conclusion and Future Works

Real-time myoelectric control requires a high level of accuracy and computational load, so a compromise between these two main factors should be considered.

The main advantage of RBF: learning takes one era, so we have a very fast network training procedure (less than a second).

The peak values functions (MAX) were extracted from facial EMG, so MAX showed the very accurate.

Also possibility of using one or two channels, and only one second of input signal were evaluated and its robustness was proved.

The proposed method of signal processing can be used for creating a robust and fast gesture classifier. Recognized signals from facial muscles can be used to control the robotic wheelchair, to control "smart house" systems and to work with a personal computer.

Also this classifier allows the developing a control system of the robotic prostheses for the disabled.

In future research for larger sample of subjects and research for new effective gesture recognition algorithms will be the development of this work. Then disabled EMG will be studied. The final stage will be a creation of a wheelchair control system by EMG.

This research is supported by Institute of nanotechnology, electronics and instrument making.

References

1. Englehart, K., Hudgins, B.: A robust, real-time control scheme for multifunction myoelectric control. IEEE Trans. Biomed. Eng. **50**(7), 848–854 (2003)
2. Hamedi, M., Rezazadeh, I.M., Firoozabadi, M.: Facial gesture recognition using two-channel bio-sensors configuration and fuzzy classifier: a pilot study. In: 2011 International Conference on Electrical, Control and Computer Engineering (INECCE), pp. 338–343. IEEE (2011)
3. Jaiyen, S., Lursinsap, C., Phimoltares, S.: A very fast neural learning for classification using only new incoming datum. IEEE Trans. Neural Netw. **21**(3), 381–392 (2010)
4. Kumaragamage, C.L., Lithgow, B.J., Moussavi, Z.: Development of an ultra low noise, miniature signal conditioning device for vestibular evoked response recordings. Biomed. Eng. Online **13**(1), 1 (2014)
5. Li, H., Zhao, G., Zhou, Y., Chen, X., Ji, Z., Wang, L.: Relationship of EMG/SMG features and muscle strength level: an exploratory study on tibialis anterior muscles during plantar-flexion among hemiplegia patients. Biomed. Eng. Online **13**(1), 1 (2014)
6. Peng, H., Long, F., Ding, C.: Feature selection based on mutual information criteria of max-dependency, max-relevance, and min-redundancy. IEEETrans. Pattern Anal. Mach. Intell. **27**(8), 1226–1238 (2005)
7. Soares, F.A., Carvalho, J.L.A., Miosso, C.J., de Andrade, M.M., da Rocha, A.F.: Motor unit action potential conduction velocity estimated from surface electromyographic signals using image processing techniques. Biomed. Eng. Online **14**(1), 1 (2015)
8. Trabuco, M.H., Costa, M.V.C., de Oliveira Nascimento, F.A.: S-EMG signal compression based on domain transformation and spectral shape dynamic bit allocation. Biomed. Eng. Online **13**(1), 1 (2014)

Proactive Robotic Systems for Effective Rescuing Sufferers

Anna Motienko[1]([⊠]), Igor Dorozhko[2], Anatoly Tarasov[2],
and Oleg Basov[1]

[1] SPIIRAS, 39, 14th Line, St. Petersburg 199178, Russia
anna.gunchenko@gmail.com, oobasov@mail.ru
[2] Mozhaisky Military Space Academy,
13, Zhdanovskaya Str., St. Petersburg 197198, Russia
doroghko-igor@ya.ru, atol-77@mail.ru

Abstract. This paper considers the task of implementing proactive control of robotic systems (RS) to rescue the sufferers. The use of a wide range of sensory elements in the RS allows you to expand the list of monitored parameters and to generate the control action with the use of predictive and proactive capabilities based on the methods and technology of integrated modeling. Consider a set of models allows to estimate the effectiveness of the existing RS rescue of sufferers or form of requirements to performance characteristics of the created RS to provide the necessary indicators of the effectiveness of rescue operations.

Keywords: Proactive control · Robotic system · Bayesian belief network · Integrated modeling · Indicators of efficiency

1 Introduction

Active implementation of automated systems in control processes and reduction of the operating personnel, as well as limited staff recruiting for emergency elimination in terms of security, require solving the problem of emergency maintenance automation to ensure the necessary level of efficiency and security. The human ambition to replace with the technical means the direct involvement in carrying out different types of work is an integral element of the civilization development process.

Recent achievements in robotics have found quite widespread use in solving automation tasks in hazardous production facilities, in performing monotonous and high-precision works. One of the problems solved by the automated control systems is failure management and prevention of contingency that cannot be solved with the implementation of reactive management, as for their implementation they require consideration of all the available statistical information and the extended control of parameters for monitoring and coordination.

In these conditions it is advisable to move to a new control technology based on the concept of proactive management, which, in general, includes, in relation to organizational and technical objects, functions of goal-setting, planning, regulation, as well functions of accounting and control, monitoring and coordination.

© Springer International Publishing Switzerland 2016
A. Ronzhin et al. (Eds.): ICR 2016, LNAI 9812, pp. 172–180, 2016.
DOI: 10.1007/978-3-319-43955-6_21

2 Concept of Proactive Management

Proactive management of objects, as opposed to traditionally used reactive management, focused on rapid response and the subsequent avoidance of possible contingencies and emergencies, involves prevention of these situations by creation in the relevant management system of fundamentally new predictive and proactive capacities at formulating and implementing control actions, based on the methods and technologies of system (integrated) modeling [1].

Proactive management methodology aims at identifying initiating events as well as forming and implementing control actions focused on countering not effects but the causes of possible contingency, emergency and crisis situations in the control object [2]. An analysis of modern automated control systems showed that they implement technical monitoring and management tools that fix the consequences of abnormal and emergency situations and provide reactive management.

In order to implement the methodology of proactive management, it is advisable to use robotic systems (RS), which have the properties of manipulation and locomotion [3, 4] as technical control means in emergency situations. Given the complexity of the accurate a priori determination of the entire spectrum and characteristics of contingency, it is necessary to develop the RS samples with the possibility of algorithmic and structural reconfiguration to eliminate unforeseen non-routine events. Such robotic systems should be built on a modular architecture, which provides the connection of a wide range of measuring instruments (sensors, cameras, technical vision systems, etc.), as well as various actuators. As an example of implementing the methodology of proactive management, let us consider the task of making a decision about assisting the sufferer of an emergency.

3 Review of Ways of Making a Decision About Assisting Sufferer of an Emergency Situation

Most researchers employ modeling and optimization to derive solutions to problems in emergency situation. Assisting the sufferer of an emergency situation include facility location, casualty transportation, relief distribution, stockpre-positioning and evacuation among many others [5].

A major part of research in extent literature focuses on transportation models alone. Horner and Widener [6], Hamedi et al. [7], Campos et al. [8], Ozdamar [9], Song et al. [10] and Barbarosoglu and Arda [11] have formulated emergency transportation models, which achieve the multiple objectives of minimizing the travel time and travel cost.

Costa et al. [12] identified the following actions that need to be developed for better performance operations: transport, storage and handling, distribution and performance evaluation. In this paper we focus on the two primary operations-casualty transportation/ evacuation and relief distribution. The major challenging areas under these primary operations are vehicle routing, network design, and location/allocation. Vehicle routing, location and allocation of various logistics problems has been the focus of many recent researches [13–15]. But most of these researches are applicable only in ordinary logistics problems. Some of researchers developed distinct models for emergency logistics operations due to its high degree of uncertainty [16–20].

4   Model of Assessing the Effectiveness of Sufferers Assistance

To inspect the disaster area it is necessary to use RS capable of detecting the injured, inspecting and questioning them as well as carrying out manipulations to determine their state, choosing the safest position for their transportation and implementing a framework for the safe casualty movement process. The peculiarity of solutions to the presented problems lies in the fact that there may be several sufferers, and when they are detected, the RS should on-site forecast the possibility of saving the sufferer.

In case of detecting a sufferer in a zone of hazards, it is necessary to assess their current state, forecast their change of state taking into account time required for assessing the current state and for their evacuation from the danger zone, and to work out a solution to carrying out the evacuation or further search for other sufferers.

To define the area to be inspected, the terrain and the effects of hazards, a model for assessing the surrounding environment is developed. Formation of the optimal trajectory for inspecting a target area and creation of the appropriate structure necessary for the transportation of the sufferer structures are performed based on the model of motion and reconfiguration of RS in different conditions. To determine the position for transporting the sufferer a model for choosing transportation position is developed. A decision about carrying out an evacuation or further search for the sufferers is made on the basis of an estimation model of the effectiveness of sufferer assistance. The interaction of given models is shown in Fig. 1.

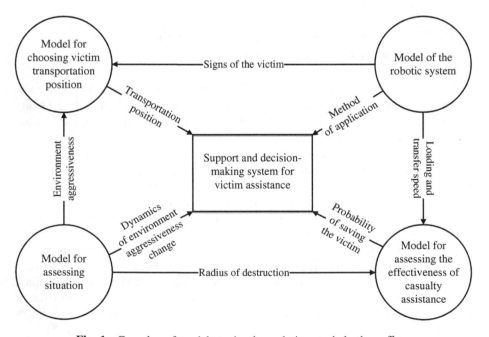

Fig. 1. Complex of models to develop solutions to help the sufferer

It is possible to identify two types of key indicators of RS application efficiency [21]. The first type of dependence of the efficiency (E) on allowed damage (D_a) is characterized by the presence of threshold damage (D^*) (Fig. 2a). The second type of dependence of efficiency on avoided damage occurs when the purpose of implementing RS is the maximum reduction of possible number of sufferers (Fig. 2b). Indicator of effectiveness is the mathematical expectation of the prevented damage $E = M(D_{prev})$.

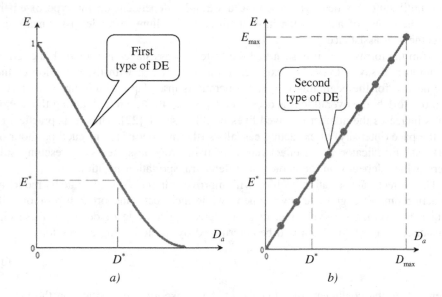

Fig. 2. Hypothetical graphics of type of dependence of efficiency (DE)

The main objective of rescue operations is detection of the maximum number of surviving sufferers and their evacuation from the danger zone. In line with this objective, the efficiency indicator is the mathematical expectation of the number of found and rescued sufferers $(N_{fr}) - E = M(N_{fr})$.

As criterion of efficiency of rescuing the sufferers K_{fr} an optimality criterion may be adopted, according to which it is necessary to detect and rescue the maximum number of sufferers per unit time:

$$K_{fr} = M[N_{fr}]/T_s \rightarrow \max, \tag{1}$$

where T_s — the time of search and rescue of sufferers.

We will take the law of death of sufferers as a basis to determine the effectiveness of rescue works:

$$N = N_0 e^{-\alpha T_s}, \tag{2}$$

where N is the number of deaths in a given time, pax; N_0 — the number of initially alive people immediately after occurrence of an emergency situation, pax; α is the rate of death:

$$\alpha = 1/T_{ms}, \tag{3}$$

where T_{ms} is the mean survival time from the moment of emergency situation occurrence until sufferer's death, representing a variable, dependent on the type of emergency, the extent of harm to the injured (in case of failure to render assistance) and correctness of assistance.

From the formula (2) it is seen that the faster the assistance is provided, the larger the number of saved lives. The experience has shown that with the reduction of the time needed for the beginning of rescue operations in collapsed buildings (structures) from 6 h to 1 h, the overall loss is decreased by $30 \div 40$ %, while doubling the rate of works increases the number of saved lives by $30 \div 40$ % [22]. Thus, it is practical to use the pace (intensity) of extracting casualties (the number of the rescued per hour of work) as an indicator of the effectiveness of using robotic systems for rescuing sufferers, which depends on the choice of sufferer transportation position.

The correct transportation position will improve sufferer's survival rate by reducing the action time of aggressive environment, while incorrect transportation position will reduce the average survival time, thereby increasing the rate of death. In these circumstances, the rate of death will be determined by the following expression:

$$\alpha = K_{tr}^{C}/T_{ms}, \tag{4}$$

where K_{tr}^{C} is the coefficient of influence of the transportation position on the rate of death:

$$\begin{cases} K_{tr}^{C} = 1, \text{if the transportation position is correct;} \\ K_{tr}^{C} > 1, \text{if the transportation position is incorrect.} \end{cases}$$

The time of search and rescue of sufferers when using RS will be determined by the following expression:

$$T_s = t_d^{ES} + t_{depl}^{RS} + t_s^{V} + t_{res}^{V}, \tag{5}$$

where t_d^{ES} is detection time of the emergency situation; t_{depl}^{RS} is time of deployment of RS; t_s^{V} is search time of the sufferer; t_{res}^{V} is time of rescue of the sufferer.

Time t_{res}^{V} comprises time t_{pos}^{V} of determining the position for transportation of the sufferer and time t_{le}^{V} of loading and evacuation of the sufferer which depend on the characteristics of RS and methods of their use.

5   Choosing the Sufferer Position with Bayesian Belief Networks

The model of choosing position for transportation of the sufferer is described in [23] and has the following form:

$$<G, PAR>, \qquad (6)$$

where **G** is the acyclic directed graph; **PAR** is a set of parameters that define a Bayesian belief network (BBN).

The vertices of an acyclic directed graph **G** are the following:

(1)   discrete variables indicating:

- transportation positions (supine position, prone position (with cushion under the chest and head), on the right side, sitting (with a raised up hand at amputation), semi-sitting position with a head bended on the chest) - $x_1, \ldots, x_5$;
- the most common injuries (injuries of the spine, fractures of the pelvis and lower extremities, brain concussion, neck injury, etc.) - $x_6, \ldots, x_{23}$;
- signs of injury (loss of consciousness, the unnatural position of the neck and back, tachycardia, headache, vomiting, speech disorder, etc.) - $x_{24}, \ldots, x_{125}$;
- methods to identify signs of trauma (examination, questioning, and manipulation) — $d_1, \ldots, d_3$ which are connected with variables $x_{24}, \ldots, x_{125}$;
- environment activity - A ("high", "medium", "low", "absent", on which depends the application of the methods of determining injury). Information about the environment activity may be entered into the robot or specified by the robot with the use of available noise or gas sensors, etc.;

(2)   the vertex of the action, indicating detection (no detection) of the sufferer, - O.

The parameters of the network PAR are the following:

(1)   for the vertex A, which does not have parental variables, a priori probabilities (unconditional probabilities) that the environment activity (enemy) is "high", "medium", "low", "absent" are $P(\tilde{A}_m)$, $m = 1\ldots4$. The sign "$\sim$" means a positive or negative definition of the variable;
(2)   for the vertices $x_{24}, \ldots, x_{125}$ conditional probabilities of dependence of signs of injury on the methods of determining injury are assigned $P(\tilde{x}_k | \tilde{d}_n)$, $k = 24\ldots125$; $n = 1\ldots3$;
(3)   for the vertices $x_6, \ldots, x_{23}$ conditional probabilities of dependence of injuries on the signs of injuries are assigned $P(\tilde{x}_j | \tilde{x}_k)$, $j = 6\ldots23$, $k = 24\ldots125$;
(4)   for the vertices $x_1, \ldots, x_5$ conditional probabilities of dependence of transportation positions on injuries are assigned $P(\tilde{x}_i | \tilde{x}_j)$, $i = 0\ldots5$, $j = 6\ldots23$.

At the initial time (the injured is not found) the robot does not perform any actions (examination, questioning, manipulation); it is assumed that the sufferer is present somewhere and has the whole set of signs of injury; the transportation position is not selected. Figure 3 shows the result of an posteriori inference in BBN with receiving the

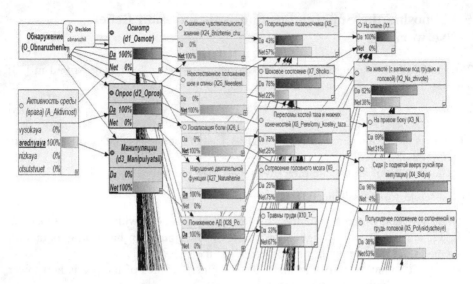

Fig. 3. Hypothetical graphics of type of dependence of efficiency

information about detection of the sufferer, environment activity – "medium" (only "Examination" and "Survey" are performed), as well as information about the sufferer.

If the sufferer is found, then the action "Examination" is carried out, and, depending on the environment activity, "Survey" and "Manipulation" are performed. As a result, some signs of injury may be excluded, and therefore probabilities of injuries associated with these signs may be reduced. This affects the choice of the position for transportation. The optimal transportation position is selected:

$$x^* = \underset{x_i \in \{x_1, \ldots, x_5\}}{\arg\max} \{P(x_i | \tilde{x}_6, \ldots, \tilde{x}_{23})\}. \tag{7}$$

Evaluation of the effectiveness of measures to rescue the sufferers showed that the number of sufferers is largely dependent on the characteristics and methods of using robotic systems. So, time t_{pos}^V of choosing the position for transportation of the sufferer can be reduced by reasonable choice of signs of injury sufficient to determine x^* (7). To reduce the feature space it is necessary to evaluate informativity indicators of signs of injuries $x_{24}, \ldots, x_{125}$.

6   Analysis of Informativity Indicators of Injuries

Currently, the following indicators are used to give quantitative evaluation of informativity of signs: ratio of dispersions of signs of standard deviation of the ensemble, values of correlation coefficients; a number of resolved disputes, distance between the centers of classes, entropy; coefficient of informativity for binary features [24, 25]. What is common to all these indicators is that they provide a ranking of features and comparative assessment of their informativity.

Given the binary nature of the signs of injury $x_{24}, \ldots, x_{125}$, in the conducted study a coefficient of informativity is used for the evaluation of their informativity [25]. Such an assessment of the informativity of signs is needed for reasonable reduction of the dimension of feature space when forming the requirements for performance characteristics of future-oriented robotic systems for rescuing sufferers [26–28].

7 Conclusion

In the event of abnormal and emergency situations, parameters of the control object or the environment interacting with it pose a serious threat to human life and health. In terms of safety, the elimination of such situations is possible only with the use of robotic systems. A robotic system comprises a sensor system (sensors, machine vision system, etc.), so it is possible to obtain additional information about the object of research, which should be used to monitor and manage coordination. For realization of these functions and transition from reactive to proactive management, it is necessary to develop a set of models and methods that allow management in view of the evolving situation. A complex of the models for decision-making about sufferer assisting emphasizes the need to predict the situation, since it is possible to spend significant resources to rescue injured people without meeting expectations. Incorrect decisions on choosing the evacuation route (without forecasting changes in the environment aggressiveness and calculating survival time) and transportation position (without evaluating a state of the sufferer) may lead to the deaths during rescue operations. This proves the correctness and feasibility of the adopted approach.

Acknowledgment. This work is partially supported by the Russian Foundation for Basic Research (grant № 16-08-00696-a).

References

1. Okhtilev, M.Y., Sokolov, B.V., Yusupov, R.M.: Theoretical and technological foundations of the concept of proactive monitoring and management of complex objects. Proc. JuFU. Tech. Sci. **1**(162), 162–174 (2015). (in Russian)
2. Okhtilev, M.Y., Mustafin, N.G., Miller, V.E., Sokolov, B.V.: The Concept of proactive control over complex objects: theoretical and technological framework. J. Instrum. Eng. **11**, 7–15 (2014)
3. Tarasov, A.G.: Prospects of creation of robotic tools and systems training and startup space rockets. H & ES Res. **6**(6), 72–75 (2014). (in Russian)
4. Tarasov, A.G., Minakov, E.P.: Robotic systems of automated control systems preparation and launching of a space rocket and indicators of efficiency of their application. Ind. Autom. Controllers **6**, 19–24 (2015). (In Russ.)
5. Caunhye, A.M., Nie, X., Pokharel, S.: Optimization models in emergency logistics: a literature review. Socio-Econ. Plan. Sci. **46**, 4–13 (2012)
6. Horner, M.W., Widener, M.J.: The effects of transportation network failure on people's accessibility to hurricane disaster relief goods: a modeling approach and application to a Florida case study. Nat. Hazards **59**, 1619–1634 (2011)

7. Hamedi, M., Haghani, A., Yang, S.: Reliable transportation of humanitarian supplies in disaster response: model and heuristic. Procedia Soc. Behav. Sci. **54**, 1205–1219 (2012)
8. Campos, V., Bandeira, R., Bandeira, A.: A method for evacuation route planning in disaster situations. Procedia Soc. Behav. Sci. **54**, 503–512 (2012)
9. Yi, W., Özdamar, L.: A dynamic logistics coordination model for evacuation and support in disaster response activities. Eur. J. Oper. Res. **179**, 1177–1193 (2007)
10. Song, R., He, S., Zhang, L.: Optimum transit operations during the emergency evacuations. J. Transp. Syst. Eng. Inf. Technol. **9**, 154–160 (2009)
11. Barbarosoglu, G., Arda, Y.: A two-stage stochastic programming framework for transportation planning in disaster response. J. Oper. Res. Soc. **55**, 43–53 (2004)
12. Costa, S.R.A., Campos, V.B.G., Bandeira, R.A.M.: Supply chains in humanitarian operations: cases and analysis. Procedia Soc. Behav. Sci. **54**, 598–607 (2012)
13. Anbuudayasankar, S.P., et al.: Modified savings heuristics and genetic algorithm for bi-objective vehicle routing problem with forced backhauls. Expert Syst. Appl. **39**(3), 2296–2305 (2012)
14. Sivakumar, P., et al.: Heuristic approach for balanced allocation problem in logistics: a comparative study. Int. J. Oper. Res. **14**, 255–270 (2012)
15. Anbuudayasankar, S.P., et al.: Unified heuristics to solve routing problem of reverse logistics in sustainable supply chain. Int. J. Syst. Sci. **4**(3), 337–351 (2010)
16. Abounacer, R., Rekik, M., Renaud, J.: An exact solution approach for multi-objective location–transportation problem for disaster response. Comput. Oper. Res. **41**, 83–93 (2014)
17. Safeer, M. et al.: A Stochastic planning model for humanitarian relief response logistics. In: International Conference on Modeling Optimization and Computing (2014)
18. Wilson, D.T., et al.: A multi-objective combinatorial model of casualty processing in major incident response. Eur. J. Oper. Res. **230**, 643–655 (2013)
19. Davis, L.B., Samanlioglu, F., Qu, X., Root, S.: Inventory planning and coordination in disaster relief efforts. Int. J. Prod. Econ. **141**, 561–573 (2013)
20. Petukhov, G.B.: Fundamentals of the theory of the effectiveness of targeted processes. Part 1. Methodology, Methods, Models (1989). (in Russian)
21. Popov, A.P., Fedoruk, V.S., Barinov, M.F., Myasnikov, D.V.: Fundamentals of modeling and evaluating the effectiveness of emergency prevention and response forces in the conduct of rescue and other emergency operations. Training Manual (2014). (in Russian)
22. Motienko, A.I., Makeev, S.M., Basov, O.O.: Analysis and modeling of the process of a choice of position for transportation of the sufferer on the basis of Bayesian belief networks. SPIIRAS Proc. **6**(43), 135–155 (2015)
23. Tarlovskiy, G.R., Fomin, Y.A.: Statistical theory of pattern recognition. Radio and Communications (1986). (in Russian)
24. Kazantsev, V.S.: Classification tasks and their software (1990) (in Russian)
25. Lbov, G.S.: Choice of an effective system of dependent features. Comput. Syst. **19**, 21–34 (1965). (in Russian)
26. Motienko, A.I., Ronzhin, A.L., Basov, O.O., Zelezny, M.: Modeling of injured position during transportation based on Bayesian belief networks. In.: Proceedings of the First International Scientific Conference "Intelligent Information Technologies for Industry" (IITI 2016). Advances in Intelligent Systems and Computing, vol. 451, pp. 81–88 (2016)
27. Kozyrenko, N.K., Meshcheryakov, R.V., Hodashinsky, I.H., Anufrieva, N.A.: Mathematical model and algorithms of people health evaluation. SPIIRAS Proc. **33**, 117–146 (2015)
28. Motienko, A.I., Ronzhin, A.L., Pavljuk, N.A.: The modern development of rescue robots, opportunities and principles of their application. Sci. Bull. NSTU **3**(60), 147–165 (2015)

Robot Soccer Team for RoboCup Humanoid KidSize League

Stepan Gomilko, Darya Zhulaeva, Dmitry Rimer, Dmitry Yakushin,
Roman Meshcheryakov, and Evgeny Shandarov[✉]

Laboratory of Robotics and Artificial Intelligence,
Tomsk State University of Control Systems and Radioelectronics (TUSUR),
Tomsk, Russian Federation
evgenyshandarov@gmail.com

Abstract. This paper describes the RoboCup Humanoid KidSize team Photon. We used Robotis DARwIn-OP robot platform and develop own software to create robot soccer team. We discuss about our software implementation of computer vision, communications, movements and main control modules.

Keywords: RoboCup · Robot soccer · Computer vision · DARwIn–OP · Humanoid robot

1 Introduction

RoboCup (The World Cup Robot Soccer) is an attempt to promote AI and robotics research by providing a task of robot soccer game for evaluation of theories, algorithms and agent architectures. In order for the robot to play a soccer game, wide range of technologies need to be integrated [1]. Today RoboCup competitions dedicated to not only soccer, but service, rescue and industrial robots.

Soccer competitions in RoboCup started in 1997 in MiddleSize, SmallSize and Simulation leagues. The Humanoid League competitions was first established in 2002 and now have become ones of the most impressive in RoboCup [2]. Humanoid League teams need to solve the problems in computer vision, locomotion, bepedal walking. Commercially available robotic platforms such as Aldebaran NAO and Robotis DARwIn-OP gave teams the opportunity to concentrate only on software development in last years [2].

Team Photon from TUSUR University participated in RoboCup 3D Simulation League in 2013–2014 and in 2015 started development in Humanoid League Soccer [3, 4].

2 Robot Hardware

As a robot platform we uses Robotis DARwIn-OP (Fig. 1). Our choice was based on DARwIn-OP open architecture, technical specification and high repairability.

© Springer International Publishing Switzerland 2016
A. Ronzhin et al. (Eds.): ICR 2016, LNAI 9812, pp. 181–188, 2016.
DOI: 10.1007/978-3-319-43955-6_22

Fig. 1. DARwIn-OP robot on a soccer field

DARwIn-OP is an anthropomorphic robot with 455 mm height and 20 DOF. Internal computer based on Intel Atom processor and works on Ubuntu Linux. DARwIn Framework includes Motion, Vision and hardware specific C++ classes.

3   Software

3.1   General Architecture

The software framework consist of four general modules: walking control, computer vision, communication and main control (Fig. 2). The software realized as modular, multi-threaded architecture. C++ and bash scripting languages using for implementation.

3.2   Walking Control Module

Walking control software module provides for robot omnidirectional adaptive movements on playing field as well as performing pre-programming movements (actions).

Pre-programmed movements (actions) required to perform routine tasks: kick the ball: left foot, right foot; get up after a fall; goal protection: one leg to the side, crouching, falling to the side of the ball and so on. This set of movements created in the RoboPlus software.

Software framework supplied with DARwIn-OP includes basic functions to ensure omnidirectional robot movement on a horizontal surface. The walking controller used allows us to tune robot gait with a large number of parameters. We used "standard" software as a base and develop only the high level functions of motion control ("moving forward", "turn on the spot", "move to the point", "follow the ball" et al.).

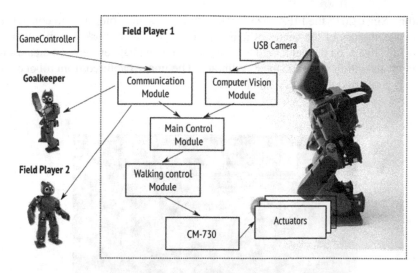

Fig. 2. Software framework

RoboCup Humanoid League Rules 2015 edition introduced a new type of field surface: artificial grass. For stable walking on this field we need to tune robot gait parameters. A lot of parameters must be tuned: increase "Foot height", change "Hip pitch offset", increase "Step right/left", reduce "Period time" and so on. This has ensured a steady movement on artificial grass field at lower speed. It should be noted that the setting walking parameters should be carried out for each robot in the team individually.

Fig. 3. Captured image

3.3 Computer Vison Module

Prior to 2015 an orange ball in RoboCup Humanoid Soccer League used. In this case we can used the center of mass method for ball finding. This is stable and high speed algorithm. Since 2015 the FIFA Size 1 ball with 50 % white color uses in competition. In this case, the use of the center of mass algorithm does not work, because there are other white objects on the field: field layout, gates, fencing banners, etc. To create a new ball search algorithm we used OpenCV library. The image captured from robot camera (Fig. 3) is converted to the HSV color space.

Fig. 4. Binarized image with detected contours

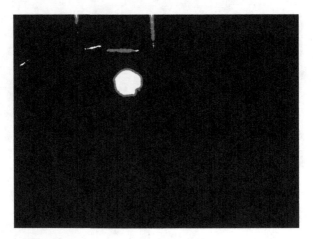

Fig. 5. Blurring image

Then the image is binarized (Fig. 4). On the resulting black and white image we detect contours. Blurring image to get rid of small objects (Fig. 5). To find ball we look for all circuits similar in shape to a circle. The ball is the large round object on the field, among the remaining contours are greatest (Fig. 6).

Fig. 6. We've got the ball!

We using histograms of oriented gradients method [5] for goal detection. The captured from the robot camera image (Fig. 7) is binarized (Fig. 8). Goal color for the current lightning conditions have been used as a binarization threshold value.

Fig. 7. Captured goal image

Binarized image is divided into 640 vertical stripes. Histogram of oriented gradients created on the number of white pixels in each vertical strip (Fig. 9). The bar chart clearly shows two peaks corresponding to goal post. Using these data, we can find the center of the goal, and robot position relative to goal.

Fig. 8. Binarized image

Fig. 9. Histogram of oriented gradients over image

3.4   Communication Module

Open source software GameController used for game control in the RoboCup Humanoid Soccer and Standard Platform leagues [6]. Communication is provided by the transfer of broadcast packets over UDP, using port 3838. The packets contains information about current match time, score, game phase et al. The game is divided into phases: INITIAL - robots cannot move, just stand; READY — must stand up to their original positions; SET — robots cannot change the position of their own;

PLAYING - the game starts; FINISHED — the game is over. The module is implemented to obtain information from GameController, send replies and set states which determines the behavior of the robot-player.

Also, communication module provides interaction between team players. Currently we are not realized the full coordination of robot activities, only small amount of information sharing between robots: estimated position, direction, distance to the ball. It is not enough to create a real game strategy, but allows, for example, autonomously make a choice of robot for ball following. Also, based on the information from the goalkeeper on the distance to the ball can be determined at which half the field is part of the game.

3.5 Main Control Module

Since the robot is a collection of a large number of both software and hardware components that work in an independent mode of each other, it was decided to build a model of management of the robot based on an event-oriented approach. Automata-based programming was used to develop the main control module.

We defined a set of states for robot on field as: FINDBALL — find the ball; BALL-FOLLOW — following the ball; FINDGOAL — search goal; KICKBALL — hit the ball; STANDUP — robot is fall, need to get up. For the goalkeeper set of states: SEARCH_BALL — find the ball; BALL_TRACKING — tracking of the ball; CATCH_BALL — implementation of protective actions; RETURN_ON_POSITION — a return to the position at the goal; STANDUP — robot is fall, need to get up.

Event markers used for change the current state: "receiving a signal from Game-Controller", "loss the ball", "ball is in the strike zone," "the robot has fallen" et al. We used different colors for robot eyes to indicate the current state.

4 Conclusion

In April 2015 the team Photon took part in Humanoid League competition for the first time on RoboCup German Open in Magdeburg and get 2nd place in the league.

References

1. Kitano, H.: RoboCup-97: Robot Soccer World Cup I. Springer, Berlin (1998)
2. Ronzhin, A.L., Stankevich, L.A., Shandarov, E.S.: International robot soccer competitions robocup and prospects of Russian teams participation. Robot. Tech. Cybern. 2(7), 24–29 (2015). in Russian
3. Gomilko, S.I., Zhulaeva, D.V., Mescheryakov, R.V., Rimer, D.I., Shandarov, E.S., Yakushin, D.O.: Robot footballers' team of robocup humanoid kidsize league. Robot. Tech. Cybern. 3(8), 11–15 (2015). in Russian
4. Gerndt, R., Seifert, D., Baltes, J., Sadeghnejad, S., Benke, S.: Humanoid robots in soccer — robots versus humans in robocup 2050. IEEE-RAS Robot. Autom. Mag. 22(3), 147–154 (2015)

5. Dalalm, N., Triggs, B.: Histograms of oriented gradients for human detection. In: 2005 IEEE Computer Society Conference on Computer Vision and Pattern Recognition, CVPR 2005, vol. 1, pp. 886–893 (2005)
6. RoboCup GameController. http://sourceforge.net/projects/robocupgc/. Accessed 5 June 2015

Semi-Markov Modelling of Commands Execution by Mobile Robot

Eugene Larkin[1(✉)], Alexey Ivutin[1], Vladislav Kotov[1], and Alexander Privalov[2]

[1] Tula State University, Tula 300012, Russia
elarkin@mail.ru, alexey.ivutin@gmail.com, vkotov@list.ru
[2] Tula State Pedagogical University, Tula 300026, Russia
privalov.61@mail.ru

Abstract. The process of control of mobile robots (MR) with typical structure, which includes sensors, onboard computer, servomechanisms being organized in multi-loop control systems, and a communicator, is investigated. It is shown, that distinctive feature of the process of MR control is rigid demands both to time of sensors/servomechanisms interrogation, and to delay between receiving information from sensors and delivering adequate command to servomechanism. Approach to modeling of the process of interpretation of external commands by onboard digital control system, based on representation of cyclogram of equipment functioning as 2-parallel semi-Markov process, is proposed. Method of transformation of 2-parallel process into ordinary semi-Markov process is worked out. Dependencies for definition of elements of semi-Markov matrix of pooled semi-Markov process are obtained. Also are obtained dependencies for calculation of time and stochastic parameters of wandering through pooled semi-Markov process and distribution of probabilities of residence of process in its states in steady state regime of functioning. Perspectives of development of direction of mobile robot control are determines.

Keywords: Mobile robot · Onboard equipment · Digital control · 2-parallel semi-Markov process · Pooled semi-Markov process · Wandering · Stochastic parameters · Time characteristics · Steady state regime

1 Introduction

Mobile robots (MR) are rather widely used for monitoring of environment in the ecological, intelligence, military and other areas [10]. In spite of dissimilarity of environments of operation (aerial, terrestrial, above-water, underwater), difference of solvable tasks, robot as object under control has the typical structure, which includes sensors, onboard computer, servomechanisms being organized in multi-loop control systems, and a communicator for receiving of outer commands and transmission of information [6,9,11].

A distinctive feature of process of MR control is rigid demands both to time of sensors/servomechanisms interrogation, and to delay between receiving information from sensors and delivering adequate command to servomechanism. Besides,

© Springer International Publishing Switzerland 2016
A. Ronzhin et al. (Eds.): ICR 2016, LNAI 9812, pp. 189–198, 2016.
DOI: 10.1007/978-3-319-43955-6_23

there is a problem of time coordination of operation of MR onboard equipment [7]. This is why evaluation of time factor on the stage of working out of mobile robot software is the important task [3].

Computer control of onboard equipment may be represented as execution a sequence of operations, every of which may be defined both by random time, being measured from the starting of operation till it finishing, and stochastic character of transition to one of the next possible operation. So, adequate approach to simulation of the MR control is semi-Markov process, which is characterized by set of states, random time of a residence in states of set, and stochastic character of switching to conjugated states. Sequence of operations forms a cyclogram of operations. Execution of external commands is reduced to changing of predetermined sequence. So problem of synthesis of cyclogram may be formulated as a problem of evaluation of time of wandering through states of semi-Markov process, and the task of definition of current state in any time moment [1,8].

2   Semi-Markov Process of Command Execution

Let us define the semi-Markov process, which simulates a cyclogram, as:

$$\mu_h = \{A, \boldsymbol{h}(t)\}, \tag{1}$$

where $A = \{a_0, a_1, \ldots, a_j, \ldots, a_{J-K}, a_{J-K+1}, \ldots, a_J\}$ – is the set of states; a_0 – is the model of the "begin" operator; $\{a_{J-K+1}, \ldots, a_J\}$ – is the subset of states, which represent the "end" operators; $\{a_1, \ldots, a_j, \ldots, a_{J-K}\}$ – is the subset of states, which represent other operators; $\boldsymbol{h}(t) = \lfloor h_{j,n}(t) \rfloor$ – is the semi-Markov matrix of size $(J+1) \times (J+1)$; $h_{j,n}(t)$ – is the element of matrix at the intersection of j-th row and n-th column.

From $\boldsymbol{h}(t)$ may be obtained the stochastic matrix $\boldsymbol{p} = \int\limits_0^\infty \boldsymbol{h}(t)dt = (p_{j,n})$ and matrix $\boldsymbol{f}(t) = \left[\frac{h_{j,n}(t)}{p_{j,n}}\right] = [f_{j,n}(t)]$ of densities of time of residence the process in its states A.

Real control algorithms are the cyclic ones, this is why in the semi-Markov model of algorithm from states of the subset $\{a_{J-K+1}, \ldots, a_J\}$ one can switching with probabilities $p_{j,0} = 1, J - K + 1 \le j \le J$ into the state a_0 during a time $f_{j,0}(t) = \delta[t], J - K + 1 \le j \le J$, where $\delta(t)$ – is the Dirac-function. If in the semi-Markov process (1) all states of subset $\{a_1, \ldots, a_j, \ldots, a_{J-K}\}$ are attainable from a_0, and from any state of subset $\{a_1, \ldots, a_j, \ldots, a_{J-K}\}$ states of subset $\{a_{J-K+1}, \ldots, a_J\}$ are attainable too, then semi-Markov process, which simulates a cyclogram of MR control, is the ergodic one.

In parallel with semi-Markov process (1) develops the process, which simulates the command generator:

$$\mu_g = \{b, [\lambda \exp(-\lambda t)]\}, \tag{2}$$

where b – is the single recurrent state; λ – is the density of flow of commands.

Process (2) is the ergodic one a-priory. Processes (1) and (2) together form the 2-parallel semi-Markov process [8], from which one should to construct the abstract united semi-Markov process, which simulates the commands execution:

$$\mu = \mu_h \tilde{\cup} \mu_g = \left\{ A', \boldsymbol{h}'(t) \right\}, \qquad (3)$$

where $\tilde{\cup}$ – is the symbol, which reflects the fact, that in (3) specific operation of pooling, which takes into account the logic of functioning of MR, take place; $A' = \{a_0, a_1, \ldots, a_j, \ldots, a_{J-K}\}$ – set of states; $\boldsymbol{h}'(t) = \lfloor h'_{j,n}(t) \rfloor$ – semi-Markov matrix of size $(J+2)(J+2)$.

Logic of functioning of MR is the next. When command in turn arrives from the generator the quasi-stochastic restart of cyclogram from states of subset $\{a_1, \ldots, a_j, \ldots, a_{J-K}\}$ through the state a_{J+1} takes place. Structure of pooled semi-Markov process constructed with taking into account the logic of pooling is shown in Fig. 1.

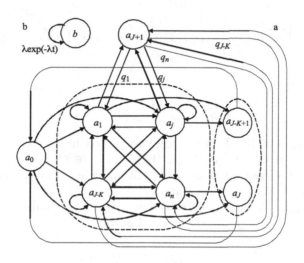

Fig. 1. Pooled semi-Markov process (a) and model of generator (b)

Additional state simulates process of cyclogram restarting after arriving the command, and provides quasi-stochastic switching mode to states of subset $\{a_1, \ldots, a_j, \ldots, a_{J-K}\}$ with probabilities $\{q_1, \ldots, q_j, \ldots, q_{J-K}\}$. In contradistinction to a_0, time density of switching to $\{a_1, \ldots, a_j, \ldots, a_{J-K}\}$ if defined by Dirac δ-function, i.e. $f_{J+1,j} = \delta[t], 1 \le j \le J - K$, due to the fact, that information about restarting state already is contained in the command from generator.

3   Stochastic and Time Characteristics of Pooled Semi-Markov Process

Let us define the elements $h'_{j,n}(t)$ of semi-Markov matrix $\boldsymbol{h}'(t)$. Let at any moment of time occurs simultaneously switching the process μ_h to state a_j and in turn switching the process μ_j. From this moment restarts simple 2-parallel process [5]:

$$
\mu_j^2 = \left\{ \{a_j, O(a_j), b\}, \begin{bmatrix} 0 & h_j(t) & 0 & 0 \\ 0 & 0 & 0 & 0 \\ 0 & 0 & 0 & \lambda \exp(-\lambda t) \\ 0 & 0 & 0 & 0 \end{bmatrix} \right\}, \tag{4}
$$

where $O(a_j)$ – is the output function of the state a_j;

$$
h_j(t) = \sum_{n=0}^{J} h_{j,n}(t). \tag{5}
$$

Between elementary processes, included into simple 2-parallel process (4) begins "competition". Density of time of switching of "competition" winner is defined as:

$$
h_j'(t) = \left[\sum_{n=0}^{J} h_{j,n}(t) \right] \exp(-\lambda t) + \lambda \exp(-\lambda t) \left[1 - \sum_{n=0}^{J} H_{j,n}(t) \right], \tag{6}
$$

where

$$
H_{j,n}(t) = \int_0^t h_j(\tau) d\tau.
$$

First summand of (6) describes the case, when in the "competition" wins process (1), and process switches into the state $a_n \in O(a_j)$. Second summand of (6) describes the case, when in "competition" wins the command generator, and process (3) switches into the state.

When switching (3) into $a_n \in O(a_j)$ density of time remaining until the next switching of command generator is defined as [5]:

$$
f_{h \to g}(t) = \frac{\eta(t) \int_0^\infty f_j \tau g(t + \tau) d\tau}{\int_0^\infty F_j t g(t) dt}. \tag{7}
$$

In accordance with results, stated in [5], if $g(t) = \lambda exp(-\lambda t)$, then $f_{h \to g}(t) = \lambda exp[-\lambda(t + \tau)]$. So restart of simple process μ^2 occurs just alike the case (4), with this difference, that (4) takes the form:

$$
\mu_n^2 = \left\{ \{a_n, O(a_n), b\}, \begin{bmatrix} 0 & h_n(t) & 0 & 0 \\ 0 & 0 & 0 & 0 \\ 0 & 0 & 0 & \lambda \exp(-\lambda t) \\ 0 & 0 & 0 & 0 \end{bmatrix} \right\}. \tag{8}
$$

After switching the process (3) into the state a_{J+1}, during the time, which is defined by degenerative distribution, with probabilities $\{q_1, \ldots, q_j, \ldots, q_{J-K}\}$ process switches into states of subset $\{a_1, \ldots, a_j, \ldots, a_{J-K}\}$.

In such a way, elements $h'_{j,n}(t)$ of semi-Markov matrix $\boldsymbol{h}'(t)$ of pooled semi-Markov process (3) are defined as:

$$
h'_{j,n}(t) = \begin{cases} h_{j,n}(t)\exp(-\lambda t), & \text{when } 0 \le j \le J, 0 \le n \le J; \\ \lambda\exp(-\lambda t)\left[1 - \sum\limits_{n=0}^{J} H_{j,n}(t)\right], & \text{when } 0 \le j \le J - K, n = J + 1; \\ q_n\delta(t), & \text{when } j \le J + 1, 1 \le n = J - K; \\ 0, & \text{in all other cases}; \end{cases}
$$

$$(9)$$

4   Properties of Pooled Semi-Markov Process

Pooled semi-Markov process (3) is the ergodic one. Thus follows from the fact that both components of pooled semi-Markov process (3) are just ergodic ones. When leaving $a_j \in A$ process return into $a_n \in A$, or switches into a_{J+1} and then return into $\{a_1, \ldots, a_j, \ldots, a_{J-K}\}$.

For ergodic semi-Markov process time and stochastic characteristics of wandering from $a_j \in A$ till $a_n \in A$ may be defined [2]. For this purpose semi-Markov the matrix $\boldsymbol{h}'(t)$ should be transformed into the matrix $^{j,n}\boldsymbol{h}'(t)$ by the next way:

– if $j \ne n$, then elements of j-th column and n-th row of matrix must be replaced with zeros;
– if $j = n$, then matrix must be increased on one row and one column, j-th column of $(J+2) \times (J+2)$ matrix must be transferred into $(J+2)$-th column, j-th column and $(J + 2)$-th row must be fulfilled with zeros.

Weighted density of time of wandering from state a_j till state a_n is defined as:

$$
\tilde{h}_{j,n}(t) = \begin{cases} \Im^{-1}\left({}^{r}_{J+1}\boldsymbol{I}_j \left(\sum\limits_{k=1}^{\infty} \left\{\Im\left[{}^{j,n}\boldsymbol{h}'(t)\right]\right\}^k \right) {}^{C}_{J+1}\boldsymbol{I}_n \right), & \text{when } j \ne n; \\ \Im^{-1}\left({}^{r}_{J+2}\boldsymbol{I}_j \left(\sum\limits_{k=1}^{\infty} \left\{\Im\left[{}^{j,n}\boldsymbol{h}'(t)\right]\right\}^k \right) {}^{C}_{J+1}\boldsymbol{I}_{J+2} \right), & \text{when } j = n, \end{cases}
$$

$$(10)$$

where $\Im^{-1}[\ldots]$ and $\Im[\ldots]$ – is the direct and inverse Fourier transform, correspondingly; ${}^{r}_{J+1}\boldsymbol{I}_j$ – is the row vector consisting of $J+1$ elements, j-th element of which is equal to one, and other elements are equal to zeros; ${}^{C}_{J+1}\boldsymbol{I}_n$ – is the column vector consisting of $J + 1$ elements, n-th element of which is equal to one, and other elements are equal to zeros; ${}^{C}_{J+2}\boldsymbol{I}_j$ – is the row vector consisting of $J + 2$ elements, j-th element of which is equal to one, and other elements are equal to zeros; ${}^{C}_{J+2}\boldsymbol{I}_{J+2}$ – is the column vector consisting of $J + 2$ elements, $(J + 2)$-th element of which is equal to one, and other elements are equal to zeros.

Due to the fact, that pooled semi-Markov process is the ergodic one, probabilities of residence in the states a_0 and $\{a_1, \ldots, a_j, \ldots, a_{J-K}\}$ may be defined as [4]

$$
\pi_j = \frac{T_j}{\theta_j},
$$

$$(11)$$

where T_j – is the expectation of time of residence in state a_j till switching into one of states $O(a_j)$ or into state a_{J+1}; θ_j – expectation of time of return into state a_j;

$$T_j = \int_0^\infty t \sum_{n=0}^{J+1} h'_{j,n}(t)dt; \tag{12}$$

$$\theta_j = \int_0^\infty t \cdot \tilde{f}_{j,j}(t)dt; \tag{13}$$

$$\tilde{f}_{j,j}(t) = \frac{\tilde{h}_{j,j}(t)}{\int_0^\infty \tilde{h}_{j,j}(t)dt}. \tag{14}$$

Let us note, that if $J - K + 1 \le j \le J + 1$, then $\pi_j = 0$. This is due to the fact that distribution of time of residence of process under investigation in mentioned states is degenerative and its density is equal to $\delta(t)$, so for this case $T_j = 0$.

Density, expectation and dispersion of time between switches are as follows:

$$f_s(t) = \sum_{j=0}^{J+1} \pi_j \sum_{n=0}^{J+1} h'_{j,n}(t); \quad T_s = \int_0^\infty t \cdot f_s(t)dt; \quad D_s = \int_0^\infty (t - T_s)^2 \cdot f_s(t)dt. \tag{15}$$

So, varying both the structure and parameters of semi-Markov process (1) Thus, by varying the structure and parameters of the semi-Markov process (1), and parameter λ of $g(t)$ and probabilities $\{q_1, \ldots, q_j, \ldots, q_{J-K}\}$, one can control of time characteristics of execution of commands by MR.

5   Experiment

The theoretical results were verified with use of Monte-Carlo method. Structure of the process of execution of commands is shown in Fig. 2.

Nodes 1, 2 and associated links simulate switching of states when absence of external commands. Probabilities of switch both from state 1 to states 1 and 2,

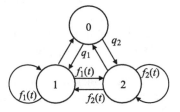

Fig. 2. Structure of process

and from state 2 to states 1 and 2. (p_{21}, p_{22}) are equal to 0,5. Densities of time of residence in states 1 and 2 are equal to

$$f_1(t) = \begin{cases} 1, & \text{when } |t - 1| \leq 0,5; \\ 0, & \text{when } |t - 1| \geq 0,5; \end{cases} \quad f_2(t) = \begin{cases} 2, & \text{when } |t - 1| \leq 0,25; \\ 0, & \text{when } |t - 1| \geq 0,25. \end{cases}$$

Density of time between two commands is the next (for transparency exponential distribution was replaced with uniform distribution):

$$g(t) = \begin{cases} 5, & \text{when } |t - 1| \leq 0,1; \\ 0, & \text{when } |t - 1| \geq 0,1. \end{cases}$$

In 2-parallel process 1 densities of time of switching of winner are the next:

$$h_1(t) = h_{w1}(t) + h_{wg1}(t),$$

where $h_{w1}(t)$ – is weighted density of switching time of 2-parallel process from the state 1 to one of states, 1 or 2; $h_{wg1}(t)$ – weighted density of switching time of 2-parallel process into state 0, which leads to further switch into state 1 or 2;

$$h_{w1}(t) = \begin{cases} 0, & \text{when } 0 \leq t \leq 0,5, \quad t > 1,1; \\ 1, & \text{when } 0,5 \leq t \leq 0,9; \\ 5,5 - 5t, & \text{when } 0,9 \leq t \leq 1,1; \end{cases}$$

$$h_{wg1}(t) = \begin{cases} 0, & \text{when } 0 \leq t \leq 0,9, \quad t > 1,1; \\ 5(1,5 - t), & \text{when } 0,9 \leq t \leq 1,1. \end{cases}$$

Probabilities p_{w1} of switching into states 1/2, or p_{wg1} of switching into state 0 are equal in the case under consideration: $p_{w1} = 0,5$, $p_{wg1} = 0,5$. Histograms for pure densities $f_{w1}(t) = \frac{h_{w1}(t)}{p_{w1}}$ and $f_{wg1}(t) = \frac{h_{wg1}(t)}{p_{wg1}}$, being obtained with use Monte-Carlo method are shown in Fig. 3.

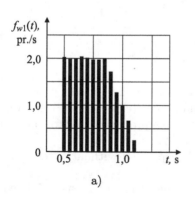

a)

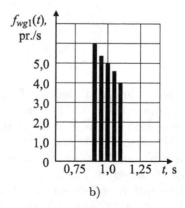

b)

Fig. 3. Densities $f_{w1}(t)$ and $f_{wg1}(t)$

Expectation and dispersion of density of time of switch 2-parallel process from 1 to 1/2 are equal 0,754 s (error 0,4 %) and 0,022 s^2 (error 0,7 %). Expectation and dispersion of density of switch time of 2-parallel process from 1 to 0 are equal 0,994 s (error 0,3 %) and 0,0007 s^2 (error 0,6 %).

In 2-parallel process 2 densities of tine of switching of winner are the next:

$$h_2(t) = h_{w2}(t) + h_{wg2}(t),$$

where

$$h_{w2}(t) = \begin{cases} 0, & \text{when } 0 \le t \le 0{,}5, \quad t > 1{,}1; \\ 2, & \text{when } 0{,}75 \le t \le 0{,}9; \\ 11 - 10t, & \text{when } 0{,}9 \le t \le 1{,}1; \end{cases}$$

$$h_{wg2}(t) = \begin{cases} 0, & \text{when } 0 \le t \le 0{,}9, \quad t > 1{,}1; \\ 5(1{,}5 - t), & \text{when } 0{,}9 \le t \le 1{,}1. \end{cases}$$

Probabilities p_{w2} of switching into states 1/2, or p_{wg2} of switching into state 0 are equal in the case under consideration: $p_{w2} = 0{,}5$, $p_{wg2} = 0{,}5$. Histograms for pure densities $f_{w2}(t) = \frac{h_{w2}(t)}{p_{w2}}$ and $f_{wg2}(t) = \frac{h_{wg2}(t)}{p_{wg2}}$, being obtained with use Monte-Carlo method are shown in Fig. 4.

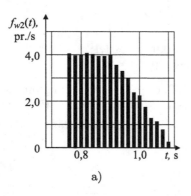
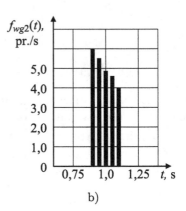

Fig. 4. Densities $f_{w2}(t)$ and $f_{wg2}(t)$

Expectation and dispersion of density of time of switch 2-parallel process from 2 to 1/2 are equal to 0,882 s (error 0,5 %) and 0,007 s^2 (error 0,8 %). Expectation and dispersion of density of switch time of 2-parallel process from 2 to 0 are equal to 0,994 s (error 0,3 %) and 0,0007 s^2 (error (0,6 %)).

Histograms of return into states 1 and 2 in process without external commands are shown in Fig. 5.

Expectation and dispersion of density of time of return in state 1 is equal to 1,98 s (error 1,0 %) and 2,104 s^2 (error 0,9 %). Expectation and dispersion of density of time of return in state 2 is equal to 2,01 s (error 0,5 %) and 2,15 s^2 (error 1,03 %).

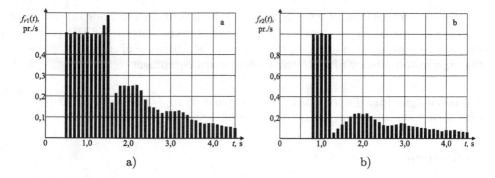

Fig. 5. Histograms of return in state 1 (a) and 2 (b) without commands

When MR is under control, which induce switches from state 0 into state 1 with probability 0,9, histogram of return into stats 1 and 2 are as it is shown in Fig. 6.

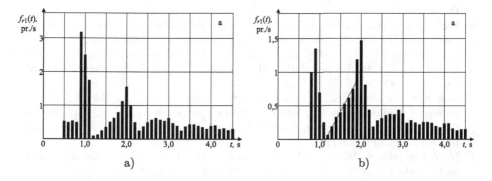

Fig. 6. Histograms of return in state 1 (a) and 2 (b) under commands

Expectation and dispersion of density of time of return in state 1 is equal to 1,27 s (error 0,9 %) and 0,339 s^2 (error 1,5 %). Expectation and dispersion of density of time of return in state 2 is equal to 2,98 s (error 0,7 %) and 2,18 s^2 (error 1,03 %).

Experimental parameters rather precisely correspond to the theoretical settlements.

6   Conclusion

The proposed model of execution of commands by MR is focused on evaluation of time intervals and current states of MR under external remote control. Dependencies obtained permit with exactness to densities to evaluate named parameters.

Further continuation of investigation may be linked with working out on the theoretical base a simple engineer methods of optimal software of MR control with use as optimization criterion the cost of delay/advance of command execution and in-consistency of functioning of onboard equipment [5].

Acknowledgments. The reported study was partially supported by RFBR and Tula Region Government, research projects No. 15–48–03232 r_a and 16–41–710160 r_a.

References

1. Breuer, H.P., Vacchini, B.: Quantum semi-Markov processes. Phys. Rev. Lett. **101**(14), 140–402 (2008)
2. Iverson, M., Ozguner, F., Follen, G.: Run-time statistical estimation of task execution times for heterogeneous distributed computing. In: Proceedings of 5th IEEE International Symposium on High Performance Distributed Computing, HPDC 1996, pp. 263–270. Institute of Electrical and Electronics Engineers (IEEE) (1996)
3. Ivutin, A., Larkin, E.: Estimation of latency in embedded real-time systems. In: 3rd Mediterranean Conference on Embedded Computing (MECO 2014), pp. 236–239. IEEE (2014)
4. Ivutin, A., Larkin, E., Kotov, V.: Established routine of swarm monitoring systems functioning. In: Tan, Y., Shi, Y., Buarque, F., Gelbukh, A., Das, S., Engelbrecht, A. (eds.) ICSI-CCI 2015. LNCS, vol. 9141, pp. 415–422. Springer, Heidelberg (2015)
5. Ivutin, A., Larkin, E.: Simulation of concurrent games. Bull. South Ural State Univ. **8**(2), 43–54 (2015)
6. Kahar, S., Sulaiman, R., Prabuwono, A.S., Ahmad, N.A., Abu Hassan, M.A.: A review of wireless technology usage for mobile robot controller. In: 2012 International Conference on System Engineering and Modeling (ICSEM 2012), vol. 34, pp. 7–12. IPCSIT (2012)
7. Li, Y.T., Malik, S., Wolfe, A.: Efficient microarchitecture modeling and path analysis for real-time software. In: Proceedings of 16th IEEE Real-Time Systems Symposium, December 1995, pp. 298–307. Institute of Electrical and Electronics Engineers (IEEE) (1995)
8. Maes, C., Netočný, K., Wynants, B.: Dynamical fluctuations for semi-Markov processes. J. Phys. A: Math. Theor. **42**(36), 365002 (2009)
9. Olsson, G., Piani, G.: Computer Systems for Automation and Control. Prentice-Hall (1992)
10. Tzafestas, S.G.: Introduction to Mobile Robot Control. Elsevier, New York (2013)
11. Wang, Z., Liang, M., Maropoulos, P.G.: High accuracy mobile robot positioning using external large volume metrology instruments. Int. J. Comput. Integr. Manuf. **24**(5), 484–492 (2011)

Smart M3-Based Robot Interaction Scenario for Coalition Work

Alexander Smirnov[1,2], Alexey Kashevnik[1,2(✉)], Sergey Mikhailov[1,2],
Mikhail Mironov[1,2], and Mikhail Petrov[1,2]

[1] SPIIRAS, St. Petersburg, Russia
{smir,alexey}@iias.spb.su
[2] ITMO University, St. Petersburg, Russia
saboteurincave@gmail.com, mironoff.togo@gmail.com,
dragon294@mail.ru

Abstract. The paper propose an interaction scenario for collaborative work of mobile robots for coalition creation and joint task solving. Scenario is considered as cyber-physical-social system that includes acting resources (mobile robots) that implements actions in physical space; information resources (robot control blocks, user mobile devices, computation services, etc.) that operate in information space; and social resources (users) that form tasks in social space. The following main operations have been identified for robot interaction scenario: mobile robot set forming that are ready to participate in scenario, coalition creation, scenario operation. Scenario has been implemented based on Smart-M3 information sharing platform that provides possibilities for different resources to share information and knowledge in cyber space with each other based on RDF ontologies.

Keywords: Mobile robots · Smart space · Interaction · Cyber-physical-social system · Smart-M3 platform

1 Introduction

Last year's a lot of research and development activities in the area of robotics have been done [1]. One of interesting and important topic in this area is information interaction support between group robots for joint task solving. It is assumed that robots have interfaces for information exchange in common shared space. In this case, it is needed to form coalitions and provide possibilities for collaborative work between robots that is significantly enhance complex tasks solving by the robots.

Cyber-physical-social systems tightly integrate physical, cyber, and social spaces based on interactions between these spaces in real time. They rely on communication, computation and control infrastructures for the three spaces with various resources as sensors, actuators, computational resources, services, humans, etc. Presented in the paper scenario is considered as cyber-physical-social system that includes three types of collaborated resources (Fig. 1):

- acting resources (mobile robots) that implements actions in physical space;

© Springer International Publishing Switzerland 2016
A. Ronzhin et al. (Eds.): ICR 2016, LNAI 9812, pp. 199–207, 2016.
DOI: 10.1007/978-3-319-43955-6_24

- information resources (robot control blocks, user mobile devices, computation services, etc.) that operate in information space;
- social resources (users) that form tasks in social space.

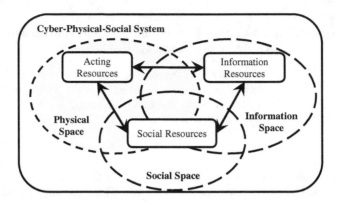

Fig. 1. Resources interaction in cyber-physical-social systems

For interaction of the cyber-physical-social system resources the smart space technology is used, which allows to provide information for sharing between different services of the system. This technology [2, 3] aims in the seamless integration of different devices by developing ubiquitous computing environments, where different services can share information with each other, make different computations and interact for joint tasks solving. In the considered approach, the main goal of smart space technology is to provide information sharing in information space of the cyber-physical-social system. The open source Smart-M3 platform [4] has been used for organization of smart space infrastructure for robots interaction. The use of this platform enables to significantly simplify further development of the system, include new information sources and services, and to make the system highly scalable. The Smart-M3 platform consists of two main parts: information agents and kernel [5]. The kernel consists of two elements: Semantic Information Broker (SIB) and information storage. Information agents are software entities, installed on mobile devices of the smart space users and other devices, which host smart space services.

The rest of the paper is structured as follows. Section 1 contains a related work analysis in the area of robot interaction systems. Mobile robot interaction scenario is presented in Sect. 2. Section 3 describes the implemented use cases. The results are summarized in Conclusion.

2   Related Work

The paper [6] considers interaction of modular robots (M-Robots) configurations or the execution of cooperative tasks such as moving objects or manipulating objects with multiple modular robot configurations. There are two types of communications are used in the approach: inter robot (between various M-Robot configurations of the colony and the control station) and intra robot (inside the M-Robot configuration). Intra robot configuration is implemented via the CAN bus technology and using master-slave architecture. For the inter robot configuration the wireless technology is used (between the control station and each M-Robot).

Authors of the paper [7] propose the mobile robotic system consisting of the mobile robot based on LEGO Mindstorms NXT kit and the OS Android-based smartphone connected to NXT controller via Bluetooth. Smartphone uses video camera and implements images analysis for determine actions to be implemented by the robot. It controls the NXT control block which controls the robot engines and sensors.

A hybrid navigation strategy for several mobile robots is proposed in the paper [8]. Authors consider a scenario with several mobile robots and obstacles in 2D space. Every robot has own task. Robots are moved in this space scan it using ultrasonic sensors and interact with each other to exchange with information about their location, movement direction, and obstacles. Proposed hybrid navigation system consists of the following three layers: high-level layer for making a global plan and a reactive algorithm runs, low-level layer for executing the plan, and deliberative layer for updating the plan in every control cycle based on the current environmental knowledge to make sure that the plan is valid and optimal.

The paper [9] describes a mobile robot based system for buildings maintenance and monitoring. The system is based on robotics integrated development environment (RIDE) platform. Every mobile robot is controlled by OS Linux-Based mini-ITX PC. The system allows different users and different robots to connect at any time. For this purposes, an identification and priority access mechanism has been proposed. The system also supports detection and recharging of low-level batteries as the robots need to be working autonomously for long periods of time.

Authors of the paper [10] mentioned that when a user and a robot share the same physical workspace the robot have to keep an updated 3D representation of the environment. Authors are used Kinect Fusion technology together with range sensors to make this representation. They are used The KinFu Large Scale (KinFu LS) project that is an open source implementation of Kinect Fusion and propose the appropriate algorithms.

3   Mobile Robots Interaction Scenario

Presented in the paper scenario for mobile robots collaboration is based on Smart-M3 information sharing platform. It is assumed that mobile robots and other resources has communication modules and can interact with each other.

3.1   Reference Model for Smart Space-Based Recourses Interaction

Reference model (Fig. 2) includes three spaces: social space, information space, and physical space. The information space consists of smart space and information services. Smart space provides possibilities for resources from social space and physical space share information with each other and with computation service.

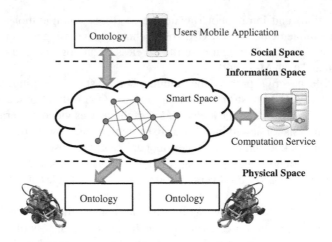

Fig. 2. Reference model for mobile robots interaction for words from letters formation

Social space consists of users who can publish and change information in smart space. Physical space consists of mobile robots.

To provide semantic interoperability between robots their interaction in smart space is based on ontologies. Each robot uploads its ontology to the smart space when it connects to the system. The ontology represents the robot. It contains information about robot requirements and possibilities. Robot requirements represent the information, which the robot needs for starting its scenario. Robot possibilities is the information that robots can provide in scope of the considered system. When a service/robot has information that can be helpful for other services/robots in the smart space, it uploads this information according to previously uploaded own ontology.

3.2 Scenario Description

The following main operations have been identified for robot interaction scenario: mobile robot set forming that are ready to participate in scenario (every robot is formed own ontological model and shares it with the smart space), coalition creation (appearing of the formalized task in the smart space, interaction of mobile robots), scenario operation (coordination, synchronization, monitoring, and regulation). It is proposed to form coalitions based on multi-level self-organization of mobile robots [11]. Robots are interact with each other and with human smartphone and decide who will participate in solving the task.

4   Use Case

Presented use case is aimed at checking the proposed reference model and coalition formation scenario. For mobile robot construction Lego Mindstorms EV3 set is used that have interfaces to interact with each other through Wi-Fi network.

4.1 General Description

A use case presented in the paper is aimed for words from letters formation by mobile robots. User forms a goal using personal smartphone. He/she determines letters sequence and correspondence between letter and colors. This information is shared with the smart space. Robots are moved in the area search letters, determine their color. Mobile robot scans the area finds a letter, determine the color and shares with a computation service current letter color. The computation service get information about current letter, letter sequence, and correspondence of colors to letters. Based on this information it calculates and shares with smart space location where this letter has to be dropped off. The robot implements a task to pick up the letter move it to specified location and dropped it off.

This use case extends previous authors work related in the area of mobile robots interaction based on Lego® Mindstorms EV3 Kit [12, 13]. Every mobile robot consists of the following main components linked together by different Lego parts (see Fig. 3): two big motors responsible for robot movement and turns; one small motor responsible both for gripping and for picking the object up; color sensor set in the gripper and responsible for the letter color determination; ultrasonic sensor responsible for the letter searching.

Fig. 3. Lego® Mindstorms EV3 Education kit and mobile robot constructed from the kit

4.2 Computation Service

For the computation service, an algorithm is designed for providing information about coordinates where a letter taken by a mobile robot has to be located. Robots able to move, search letters using onboard sensors, and pick them up and drop them off. Every letter has the unique color and robots are used a special color sensor to determine the letter. A robot shares information about found letter with smart space and then it is processed by computation service to get coordinates of the place where the letter has to be dropped off. The flowchart of the designed algorithm is shown in Fig. 4. The computation service joins smart space and subscribes to the following triple:

(Null, "holdsColor", Null).

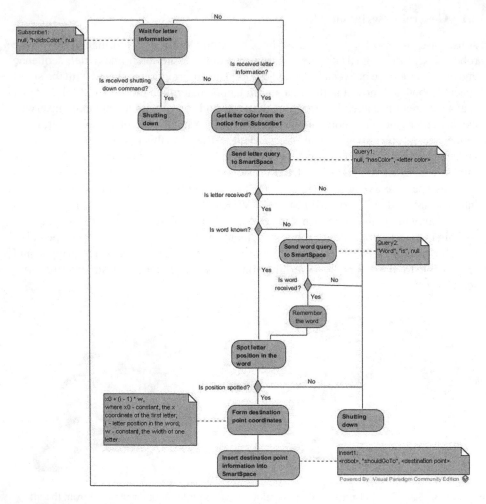

Fig. 4. Algorithm flowchart for coordinates providing to robots based on found letter

When the triples with the predicate "holdsColor" are inserted into smart space by a robot, the computation service gets appropriate notification (triple that includes name of a robot that finds a letter and the letter color). After receiving information about the letter color the service make the following queries to determine a sequence of the letter in the word:

(Null, "hasColor", [letter color]).
("Word", "is", Null).

After the letter position determination the computation service calculates and forms the coordinates of the point to the letter drop off. Abscissa in centimeters is calculated using the following formula:

$$x = x_0 + (i - 1)w,$$

where x_0 – constant, x – coordinate of the letter, in centimeters, i – letter position in the word, w – constant (width of one letter in centimeters). Ordinate is a predefined in advance constant that is a middle of the table where robots are moved. Then the computation service shares with smart space the following triple:

("Robot", "shouldGoTo", [destination point]).

4.3 Mobile Robot Control Service

For every robot in the considered scenario the following service has been developed that is uploaded to participating robots controllers. In the propose use case robots have to recognize letter colors. It is implemented by special light sensor (included in the Lego Mindstorms EV3 kit). Scenario has been implemented with assumption that letters are located on one side of a robot. Robots are moving forward and scan area with an ultrasound sensor. As a result, we have a data array which compares a distance traveled by the robot and ultrasonic sensor's measurement.

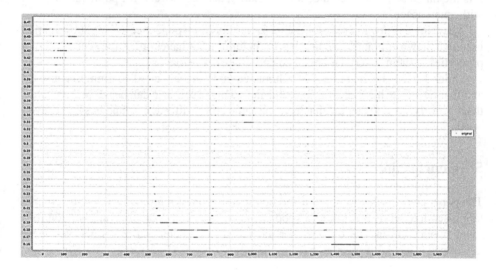

Fig. 5. Using ultrasonic sensor for finding letters (example for letter "M")

In Fig. 5 the abscissa is represented by motor's turns count which increased with the robot movement. The ordinate is represented by distance acquired by the ultrasonic sensor from the robot to the objects. The smaller value on the axis means the closer object to the robot. The main goal is determination the center of the object according to the values from the ultrasonic sensor. For achievement of this goal, we have to filter the data for an object's selection and working on them. At this moment we use cut the upper half of graph ordinate because it allows us to accurately determine the boundaries of the object and the distance between objects. After filtration, we have to calculate center of the letter for further transporting. Upper layer of the rest of data was taken for letter's shape localization and calculation of the letter center as the arithmetic average of border's values. If we can't define the boundaries of the object (ultrasonic sensor adds noise to measurement or filter cut one of borders of the objects, e.g.), then we take another layer of data and repeat calculation again until the center of letter is not found.

4.4   User Mobile Application for Collaboration with Robots

For providing the user convenient interface for goal-setting the mobile application has been developed. Mobile application is intended to provide possibility for the user to set/ change a sequence of letters and color of every letter and publish this information with smart space. A user-friendly graphical interface has been developed using of Java programming language. Application has been developed based on Android Studio Integrated Development Environment (IDE), that officially recommended by Google. Beside it, the IDE that based on IntelliJ IDEA engine have comfortable interface settings application color palette and built-in mechanisms for working with version control systems. Application has three visible activities that has identity color palette. First of it is the initial screen (for connection to smart space) and screen for publishing a word with smart space.

The third activity is screen for colors for letters correspondence setting. Information is shown in two columns. The first has Latin letter in alphabet order, the second – color properties of them. Every letter can be one of the six colors or none.

Color displays letters in the first column depends on the second column. Because the application palette has white background and colors in scenario can be white the method that correctly show the letters that have the same color with background has been developed.

5   Conclusion

The paper presents the description and implementation of reference model, the computation service, services for mobile robots control, and the user mobile application for a word publication in smart space. For the scenario implementation letters have been developed using 3D printer. Robots are based on Lego® Mindstorms EV3 Kit.

Acknowledgements.   The presented results are part of the research carried out within the project funded by grants # 16-07-00462, 16-29-04349 of the Russian Foundation for Basic Research, programs # I.5, III.3, and # I.31 of the Russian Academy of Sciences. The work has been partially financially supported by Government of Russian Federation, Grant 074-U01.

References

1. Grieco, L.A., Rizzo, A., Colucci, S., Sicari, S., Piro, G., Di Paola, D., Boggi, G.: IoT-aided robotics applications: technological implications, target domains and open issues. Comput. Commun. **54**, 32–47 (2014)

2. Cook, D.J., Das, S.K.: How smart are our environments? an updated look at the state of the art. Pervasive Mob. Comput. **3**(2), 53–73 (2007)
3. Balandin, S., Waris, H.: Key properties in the development of smart spaces. In: Stephanidis, C. (ed.) UAHCI 2009, Part II. LNCS, vol. 5615, pp. 3–12. Springer, Heidelberg (2009)
4. Smart-M3 at Sourceforge. http://sourceforge.net/projects/smart-m3
5. Honkola, J., Laine, H., Brown, R., Tyrkko, O.: Smart-M3 information sharing platform. In: Proceedings of ISCC 2010, pp. 1041–1046. IEEE Computer Society, June 2010
6. Baca, J., Pagala, P., Rossi, C., Ferre, M.: Modular robot systems towards the execution of cooperative tasks in large facilities. Robot. Auton. Syst. **66**, 159–174 (2015)
7. Ono, K., Ogawa, H.: Personal robot using android smartphone. Procedia Technol. **18**, 37–41 (2014)
8. Zhu, Y., Zhang, T., Song, J., Li, X.: A hybrid navigation strategy for multiple mobile robots. Robot. Comput. Integr. Manuf. **29**, 129–141 (2013)
9. López, J., Pérez, D., Paz, E., Santana, A.: WatchBot: a building maintenance and surveillance system based on autonomous robots. Robot. Auton. Syst. **61**, 1559–1571 (2013)
10. Monica, R., Aleotti, J., Caselli, S.: A KinFu based approach for robot spatial attention and view planning. Robot. Auton. Syst. **75**, 627–640 (2016)
11. Smirnov, A., Kashevnik, A., Shilov, N.: Cyber-physical-social system self-organization: ontology-based multi-level approach and case study. In: Ninth IEEE International Conference on Self-Adaptive and Self-Organizing Systems, Cambridge, MA, pp. 168–169 (2015)
12. Smirnov A., Kashevnik, A., Teslya, N., Mikhailov, S., Shabaev, A.: Smart-M3-based robots self-organization in pick-and-place system. In: Proceedings of the 17th Conference of the Open Innovations Association FRUCT, Yaroslavl, Russia, pp. 210–215 (2015)
13. Smirnov, A., Kashevnik, A., Mikhailov, S., Mironov, M., Baraniuc, O.: Multi-level robots self-organization in smart space: approach and case study. In: Balandin, S., Andreev, S., Koucheryavy, Y. (eds.) NEW2AN/ruSMART 2015. LNCS, vol. 9247, pp. 68–79. Springer, Heidelberg (2015)

Speech Dialog as a Part of Interactive "Human-Machine" Systems

Rodmonga Potapova[✉]

Institute of Applied and Mathematical Linguistics,
Moscow State Linguistic University, Ostozhenka 38, Moscow 119034, Russia
rkpotapova@yandex.ru

Abstract. One of the most significant features in applied linguistics and recognition technologies used in methods of spoken language recognition is speech signal which includes some primary tasks: preprocessing, processing and recognition regarding next main important features of acoustic analysis of spoken language: F_{0i}, I_i, t_i, F_{ni}. This paper presents one of the human machine methods with regard to continuous speech detection on the basis of formant Fni analysis. There are many ways to perform acoustic analysis, but the acoustic-phonetic recognition functions at the phoneme and prosody level seem to be one of the classical speech recognition methods.

Keywords: Speech recognition · Dialog system · Acoustic features · Formant tracking analysis · Prosody

1 Introduction

The study of the spoken language role in communication between human and machine and the development of automated systems with languages of communication natural for humans are under development. However, it has been already established that the allocation of similar functions between human and machine must be dynamic in the process of interaction based on certain performance criteria (e.g., solution time, costs, etc.) [1], [7,8], [12]. Considering the human-machine interaction, let us agree to consider a dialog as a time management mode, in which there is at least one of the following feature present: aim of the multimodal interaction for both partners, a certain degree of equivalence of their activities in the process of solving a problem, expansion and improvement of knowledge (skills) of one partner based on the knowledge base (skills) of the other partner, particularly one partner learning from the other one, messaging aimed at mutual understanding. Requirements for the dialog organization are as follows: ease of learning in the use of dialog means, good error detection, efficiency, consistency, adaptability, effectiveness, etc. [2,3], [5], [9], [13], [15].

One of the forms of human and machine interaction, which is considered promising and is intensively developed in recent years, is a speech dialog as a part of interactive human-machine systems [14]. The advantages of spoken

© Springer International Publishing Switzerland 2016
A. Ronzhin et al. (Eds.): ICR 2016, LNAI 9812, pp. 208–218, 2016.
DOI: 10.1007/978-3-319-43955-6_25

speech over all other means of communication are as follows: hands are free; it is easy to give special signs; no tools are required; the operator is not to take a fixed position; lighting conditions and mechanical vibration are of no importance, etc.; no keyboard and display needed; simple method of observing the response to the command; one can use a common channel for several purposes; the human-machine system may include people with disabilities; protection against unauthorized access is allowed by means of voice recognition; it is easy to provide interface to telephone systems.

However, voice communication has its drawbacks, which include exposure to noise disturbances; difficulty in singling out random inclusions in large volumes of information; inability to have unlimited data input into a machine for a long time. Using natural language in human-machine interaction implies knowledge of the specifics of its structure [4], [11]. First of all, an approach to language as a tiered phenomenon is widely used. A linguistic tier (level in linguistics) refers to one of the way to consider the language determined by the properties of units singled out by a researcher. Usually the following tiers are identified: phonetic, phonological, morphological, lexical, syntactic, and semantic one [9]. The term level is widely used in linguistics. As this term is also used in communications technology, but with a different meaning (e.g., sound pressure level, volume level, etc.), and the term quantization level is used in digital technology, in order to avoid confusion, in this paper the term level is replaced by the term tier.

The smallest unit of the peripheral tier of the language is a phoneme, which serves to recognize words and their forms. For each language, there is a limited set of vowels and consonants. When writing words phonemes are conveyed by corresponding graphemes. Implementation of phonemes in a speech flow depending on its position in a syllable or word, on a specific sound environment, etc. is called an allophone (variant) of this phoneme [6,7], [10]. Rules of phonemic classification and combinatorics in language are studied in phonology (in particular, one of its sections, phonotactics). A phonetic tier relates to the study and description of specific features of implementation (articulatory, acoustic, auditory) of phonemes in the speech flow. Each phoneme is characterized by a set of different (differential) features based on the description of articulation (speech production) and acoustic properties of the phoneme. Such features include, for example, deafness (no pitch is present), voicing (presence of pitch).

A morpheme is a unit of a higher tier than a phoneme. Using morphemes one can describe the word structure consisting usually of a root morpheme (root), prefixal and suffixal morphemes (prefixes, suffixes). The lexical tier covers the vocabulary (words and word forms) of a language. The syntactic tier includes such language units as a phrase, sentence. Using the syntax rules of construction of phrases and sentences can be described and formalized for each language. The semantic tier may include meanings of words and various grammatical forms. Using the semantics, a relationship is established between objects of reality (the real world) and their linguistic symbols that is, words.

Using the pragmatic tier a relationship is established between language units and the user. Rules for constructing words and their changes, and structures of phrases and sentences are described in a special linguistics section, grammar.

In recent years, so-called context-free grammars become widely used; these are mathematical models of the syntactic structure of a sentence. An example of such a model is a tree (network) model, or a graph model. A set M with a given binary relation R is called a graph and is denoted as <M; R>. Elements of M are called nodes of the graph; the finite graph is called a tree. Any sentence can be represented as a finite linearly ordered set. As part of the phonetic tier, the following is studied: acoustic characteristics of speech, features of auditory perception, articulation, i.e., the process of speech production and coarticulation the process of articulatory mutual interference of sounds in the speech flow. As a rule, coarticulation is most pronounced within the shortest unit of pronunciation, a syllable. As a result of coarticulation there is a mutual assimilation of articulatory movements of neighboring sounds, and, consequently, their acoustic features [9].

2   Research and Discussion

Sensory robotic devices are defined as devices equipped with simulators of human senses, i.e. as sensitized systems. Locomotion robotic systems are characterized by active movements in space. Interactive robotic systems involve systems that interact with each other and with the operator, who, in particular, uses voice control. The bionic approach to the development of robotic systems implies the use of bionics data, which studies features of organisms structure and activity to create new tools, mechanisms and systems. A special role is played by the use of the association principle of constructing a model of central speech processing mechanisms, which is based on biological facts about the structure and properties of neurons and their associations [4].

During processing of spoken speech, systems of speech automatic understanding are mostly used. In contrast to speech recognition, their task is to establish the meaning, i.e. semantic interpretation. Such systems use additional information widely to establish the grammatical or semantic correctness of the utterance (sentence). Using these systems one correctly evaluate and even supplement poorly recognized parts of the sentence, so it can be said that these systems are able to intelligently guess.

The main difficulty both for recognition and for understanding of speech lies in identification of reliable signs at the acoustic-phonetic tier. Although speech is a signal that varies over time, at the first stage stationary portions of segments can be studied in order to identify distinctive acoustic parameters. The study, which was intended for searching formants and their values led to the following results: accurate extraction of formants is not entirely a reliable operation; preparation of phonetically correct classification of the sound spectrum is possible even with the help of a schematic (rough) description of acoustic characteristics; accuracy of the spectral analysis is redundant.

In a series of studies a program was designed and experimentally tested to recognize individual words using a limited number of frequency parameters and focusing on a description of temporal changes. Characteristics obtained as a result of distinguishing similar words are rather significant in terms of validation of parameters used, although it refers mainly to experiments with one speaker. The analysis of natural speech samples when using segments of various lengths revealed that the segment duration of 5 ms is optimal. All these observations have been taken into account in the development of a reliable algorithm for formant extraction.

The *formant tracking algorithm* is used quite successfully, but it does not give a smooth trajectory as formant frequencies obtained have some errors that occur for various reasons. This primarily relates to the beginning of the explosion, where the signal level is quite low, and the spectrum changes rapidly. The speed of formant changes is usually the highest at the beginning of the explosion, and then tends to zero gradually in the areas of further transition, so the exponential model is used to calculate the formant frequency:

$$F_i(t) = F_{ai} + (F_{bi} - F_{ai})[1 - exp\{-(t - t_0)/\tau\}] \text{ and } (t - t_0),$$

where F_{ai} is the initial value of F_i in the explosion segment (formant locus); F_{bi} is the finite value; t_0 is the beginning of the explosion; τ is time constant; (t) is a function of the temporal segmentation.

When a number of formant frequencies is set on a limited transition area, all these parameters except t_0 can be identified uniquely with a minimum mean square error. The above approach can provide a good approximation in detection of the actual trajectory of the formant only when a corresponding temporal interval of the trajectory is given. For a more precise formant extraction data in the field of noise components should be excluded due to large individual differences in the duration of the noise. Since the noise segment for voiced plosives is characterized by a predominance of low-frequency components and the overall low energy, speech information within the relevant frequency range can be useful for automatic detection of this segment.

Our data indicate that formant loci of voiced plosive consonants are directly dependent on the context, i.e. vary greatly due to coarticulation, especially for consonants b and g, even if their areas are divided into F_2 and F_3 planes when data obtained is processed independently of the context of the vowel. One might also notice that neither their relative position nor their relationship with the stationary (central) portion of the subsequent vowel does not remain invariant, hence recognition of these consonants must be based on recognition of the context of the vowel. Studying the possibility of using temporal characteristics indicates that the time of energy increase could significantly increase the degree of separation between voiced plosives, particularly between consonants b and g [9].

The process of integration of information on the parameters and attributes has not been sufficiently studied. One possible explanation is proposed in a model, according to which each of the attributes with a sufficient degree of certainty is identified as belonging to one of several classes. Then the process of collation takes place, during which the identification results for each phoneme are compared with output results by feature detectors. Errors are characterized primarily by presence of omissions due to the following factors: at the transition between two sounds it is difficult to separate them. It is particularly difficult to separate semivowels, liquid and nasal sounds from vowels nearby, which leads to phoneme skip error; sometimes explosion after a voiceless closure at the beginning of a word takes place so quickly that it is difficult to define the class of phonemes; it happens sometimes that certain distinctive features are under-represented by their acoustic keys.

So, a fricative consonant can be taken for a closure preceding an explosive consonant; in connected speech (even within a single word) some sounds are much reduced (and even dropped) that relates the problem of the ideal phonetic coding of the vocabulary. This raises the question of multiple representations of the vocabulary. In addition, errors occur due to repetition and insertions: sometimes within such groups of phonemes as *dr, gl, gr, pl* where a vowel insertion between two consonants is found; it is believed that one phoneme is segmented into two phonemes, e.g. vowels *i* and *u* followed by consonant *r* often yield two vowels; in connected speech in the presence of following vowels the program can detect additional phonemes between them.

Modelling of possible errors is constructed based on false or correct segmentation; *presence of insertions; omissions; segment merging; decomposition of segments into additional segments.* A method is known that serves to interpret speech patterns containing explosive consonants. Interpretation of speech patterns includes the construction of hypotheses about possible phonetic transcription of syllabic segments that are automatically allocated from the digital representation of energy - frequency - time obtained from the instantaneous spectral analysis of the voice message. Each hypothesis is evaluated, and a certain degree of reliability is attributed to it, which allows further evaluation of the hypotheses regarding the words, syntactic structure and semantics of the spoken sentences.

The approach and rules described in a number of papers are an attempt to formalize the intuitive logic used by phoneticians. Numeric functions are defined as a probability measure of the speech pattern interpretation. The advantage of this model is its flexibility. One can enter any ratio and any restriction obtained in the course of research in the field of phonetics, and thereby optimize the recognition performance. The algorithms used in some speech recognition systems have several tiers and sources of knowledge. Each source of knowledge contains syntax rules that correlate a unit of this tier represented by a syntactic category with a unit of the lower tiers. Attempts have been made to obtain rules that explain context-related dependencies to resulting from coarticulation, using automatically identified characteristics.

The speech signal at the input is subjected to an analysis based on the fast Fourier transform. Then, spectra processing is performed to obtain the following characteristics: S - total spectrum energy; B - energy in the frequency band of 200 ... 900 Hz; F - energy in the frequency band of 5 ... 10 kHz; A - energy in the frequency band of 3 ... 5 kHz; R_v - relationship between B and F. Phonetic characteristics are obtained using a graph dictionary. The context specificity of the rules is based on the articulation place of the preceding and subsequent vowels. The recognition rules for stop consonants are based on information about the formant transitions and descriptions of explosion spectra. To obtain speaker-independent rules, one should consider not absolute values of measurements, but relative ones.

The applied algorithm is based on information about the phonetic characteristics. It makes it possible to accurately determine the beginning and end of formant transitions, avoid measuring F_2 and F_3 during the closure, noise interval and explosion. Conversion of acoustic data into phonetic and phonemic ones is performed by a system, which puts forward and tests hypotheses on various tiers. The following acoustic features are identified and used for pre-category classification: ratio of low (200 ... 900 Hz) and high (5 ... 10 kHz) frequencies of Rv; Rv parameter dispersion; sound duration; dispersion of total signal S energy; ratio of S and R_v dispersion values. It should be noted that the representation of the vocalic context is based only on the articulatory description of vowels, for example, the place of their articulation determined by values of F_2 on the stationary part of the vocalic segment.

The research results indicate that by introducing coarticulation restrictions in the recognition algorithm better results can be obtained than with methods previously used. This approach can be used to distinguish various voiced consonants in positions between vowels. In speech recognition a number of parameters are used, most of which have theoretical equivalents and contain an equivalent amount of information. The functions of the vocal tract cross-sectional area seem to be the most preferred parameters, as they correlate most to movements of the tongue, lips, and other organs of articulation. In recognition of words the main problem is lexical search, i.e. search for the right words.

Two recognition methods have been evaluated based exclusively on acoustic information. The first one is that the vocabulary is viewed sequentially and for each word a similarity matrix is calculated, i.e. a list of first candidates is formed. Then, an algorithm of sequential decoding is used. However, this method is impractical since the average task time is 18 s for a single word. To reduce time for lexical search, the second method is used the method of multi-tier approach with silhouettes. The vocabulary is represented on various tiers lists by lists of silhouettes. These tiers are obtained by means of phonetic coding based on increasing accuracy. For example, vowels come first, then consonant phonemes follow, further an oral (buccal) vowel is refined, then a nasal vowel, a consonant semivowel, etc. Every word is discarded for which no silhouettes are among the first candidates. This processing ends as soon as the number of candidates becomes equal to the specified one. Then, the more accurate recognition is performed relying on the phonetic transcription for each word. Noteworthy is also the method for best matching between the classified phonemic chain and a phonemic chain of a lexical unit in the vocabulary. Here, a phonemic similarity matrix and LP are used.

The use of prosodic information implies reference to other tiers of the language (syntactic, semantic ones), which provides a relatively self-contained method of acoustic detection of syntactic structures, independent from potentially erroneous sequences of hypotheses of the words derived from the input acoustic information [6,7], [10]. Modern scholars expand the main prosodic features helping to implement the automatic recognition of continuous speech, including the stress, the F_0 envelope of the utterance, rhythm, pausation, tempo, change in the energy configuration and temporal correlation of segments in pre-juncture and post-juncture positions at the boundary of sentences and syntagmas. The task of automatic continuous speech recognition cannot be accomplished without the inclusion of information on macrosegmentation of the text into phrases and minimal semantic fragments extracted from the speech flow syntagmas.

The development of the program for macrosegmentation of continuous speech based on prosodic, syntactic and semantic information requires a series of studies aimed at solving the following tasks: identification of prosodic characteristics of continuous speech segmentation on the hearing level; definition of the role (functional weight) of individual prosodic parameters when marking boundaries in continuous speech macrosegmentation; analysis of the effect of the rhythmic structure of pre- and post-juncture words; syllable type (stressed, not stressed, those in absolutely final positions, non-final positions, etc. ...) and the phonetic quality of a stressed vowel on prosodic characteristics of pre- and post-juncture phonetic words at the boundary of syntagmas and phrases. Since prosodic information can influence the results of the recognition system, it is assumed that a source of knowledge about prosodic information is needed that would be a module of the recognition system, which can generate hypotheses or produce other knowledge. Since almost all the prosodic algorithms are based on accurate input information extracted from the acoustic signal, the great emphasis is put on the development and organization of prosodic parameter extraction algorithms. It is worth noting the possible impact of prosodic information on the higher parts of the speech understanding system. The fact that such influence exists is well known from works on human perception of speech, but it has not been considered at all or has been not sufficiently taken into account in speech recognition systems.

There is a well-known method of interpretation and prosodic stylization. This method allows, on the one hand, to identify the boundaries of prosodic groups and syntactic constituents and, on the other hand, to establish the hierarchy of semantic utterance constituents. The basic principles of this method are as follows: (1) prosody is a multiparameter structure, selection of a single parameter, usually F_0, can lead to errors in the interpretation, as changes in F_0 may reflect various functions and not be in direct relation with the syntactic and/or semantic organization of the utterance; (2) extraction of prosodic invariants is related to the detection procedure involving transformation of objective data. This transformation allows erasing microvariations and conversion of perceptual data. It is important to properly distribute the relations between prosody and accent, syntax and semantics from the very beginning. Prosody has a demarcation function of the syntactic constituents. It also has a semantic function in the organization of the utterance. To avoid errors, the pitch raise and fall contour should be calculated cyclically from the first detected maximum value.

In phonetics, the standpoint regarding the acoustic emphasis of the phonetic word boundaries (rhythmic structure, RS) has undergone a number of changes. The complete denial of the acoustic boundaries of a phonetic word was replaced by statements that, in determining the phonetic word boundaries in the speech flow it is quite practicable to rely on objective criteria: acoustic characteristics of sounds at the juncture of phonetic words and their allophonic variance. In delimitating the speech flow into phonetic words (and RS, too), acoustic characteristics of juncture sounds is required in all cases: both without and with a pause. As shown by a series of studies, the probability of a speech pause is dependent on the following: nature of sound combinations (it is known, for example, that a pause occurs in 95 % of cases in CV (consonant-vowel) combinations and in 5 % of cases in VC combinations); rhythmic structure of neighboring words (for example, if the first word ends with a stressed syllable, and the next word begins with a stressed vowel as well, a pause between these words is more likely than in the case where the stressed syllable of the first phonetic word is followed by an unstressed syllable of the second phonetic word); place of the considered juncture in the phrase. Along with the above linguistic factors that condition a pause between words, there are non-linguistic factors such as an individual style of pronunciation, breathing in the process of speech production, etc. A pause serving as a boundary signal helps to identify the juncture sounds. As shown by spectral and prosodic studies the mutual influence of juncture sounds in words separated by a pause is observed in a number of cases, for example in a sequence of a labial vowel and a consonant separated by a pause. The presence of a pause of a certain duration in this case does not break coarticulation implemented in this sound sequence. Of all the prosodic characteristics the most indicative one for phonetic word segmentation is the duration.

Conclusions on the duration depending on various types of junctures are obtained from a material of various languages, including Russian, German, English, etc. If one assumes that the intensity as compared with other physical characteristics better determines the word structure forming a power arc, one can say that intensity variations at the juncture of words also indicate a word boundary. The determination of phonetic word boundaries in the speech flow is accompanied with a number of difficulties arising due to the following factors: phrase belonging to the style of pronunciation and type of utterance; position of the phonetic word in the text; position of the phonetic word in the syntagma and phrase. In the Russian material the boundaries of a phonetic word in the speech flow are determined by the following: absence of vowel reduction of the absolute beginning of the phonetic word (with the exception of the position after an unvoiced consonant); devocalization of voiced consonants at the absolute end of words in positions before vowels and sonants; formation of a special kind of a geminate (doubling) at the juncture of consonants with the same phonetic quality. In determining the boundaries of a phonetic word in the speech flow it is necessary to consider all these factors, as well as to correlate the frequency envelope of the pitch and intensity with changes over time. The acoustic characteristics in various combinations in the speech flow form acoustic parameters of boundary signals.

The study of prosodic parameters of the spoken text goes beyond theoretical issues and is of great importance for practical purposes: development of algorithms that provide automatic verbal human-computer communication. At the same time a number of difficulties arise related to the ambiguity of the speech signal, and perculairities of continuous speech segmentation. At the present stage the removal of ambiguity in acoustic terms is achieved by including control units in the system using the information of higher tiers of language. However, the lack of knowledge in the field of acoustic features of the boundaries of semantic fragments of the text makes it difficult to fully use such sources of information as the syntax and semantics, for any system that claims to properly recognize and understand speech, should find segments necessary for subsequent operations following certain features, distinguish essential features of these segments from non-essential ones, use additional cues that are stored on the periphery of knowledge, etc. Finding reliable prosodic features of semantic segmentation of the text is of particular importance, as it makes it possible to perform proper semantic segmentation of the spoken text, to identify the inventory of resulting fragments of the spoken text, to determine types of relationships between semantic fragments of the text. In this regard, it seems promising to use a search strategy, which includes the study of prosodic feature modifications on the basis of two types of speech implementations per word and continuous (within the text) and the identification of prosodic features providing reliable segmentation of the spoken text.

3   Conclusion

Continuous improvement of dialog communication forms between a human operator and a machine should lead to optimization of the dialog between them. The human machine dialog in natural language involves the use of relevant technical methods and specific linguistic knowledge. To develop such a dialog with a view to the creation of next generation computers requires efforts of specialists in different fields of science. Studies have shown that close cooperation with linguists is needed, as currently under developed projects of speech-controlled automated systems are beyond the capability of one group of scientists. It is known that at this stage the linguistic level of the human machine dialog is still quite low, and it makes it difficult to create next-generation computers and impedes development of optimal speech robot control. Researches in the field of human machine interaction proved the necessity of experiments for the study of learning processes with the use of machines, formalization of natural language, modeling human machine dialogs in natural language, development of machine concepts, expert systems, databases, and knowledge bases.

The term dialog usually refers to the process of direct communication between two subjects, in which there is constant change of speaker and listener roles.

A dialog between people implies, as a rule, the presence of targeted messaging, mutual understanding of partners, a certain equivalence of all their activities in the process of messaging, enhancement of their knowledge and skills. A similar definition is used both when considering the verbal interaction between people and in construction of a dialog between a human and a machine. However, there are other approaches to the interpretation of a human machine dialog. The most common interpretation is as follows: human and machine perform dynamically changing functions, thereby increasing the efficiency of the whole process of solving a problem, starting with its wording and ending with the execution of a debugged program. Mutual understanding is manifested in each of the partners knowledge of the system of linguistic signs or codes, which build individual messages, as well as in the presence of at least partially coinciding views on the topic of their conversation.

Acknowledgments. The research was financially supported by the Russian Foundation for Basic Research, grant No. 14-06-00363.

References

1. Anzalone, S.M., Yoshikawa, Y., Ishiguro, H., Menegatti, E., Pagello, E., Sorbello, R.: Towards partners profiling in human robot interaction contexts. In: Noda, I., Ando, N., Brugali, D., Kuffner, J.J. (eds.) SIMPAR 2012. LNCS, vol. 7628, pp. 4–15. Springer, Heidelberg (2012)
2. Bertau, M.-C.: Voice as heuristic device to integrate biological and social sciences. A comment to Sidtis & Kreimans in the beginning was the familiar voice. Integr. Psychol. Behav. Sci. **46**(2), 160–171 (2012)
3. Khnel, C.: Introduction and motivation. In: Quantifying Quality Aspects of Multimodal Interactive Systems. Part of the series T-Labs Series in Telecommunication Services, pp. 1–11 (2012)
4. Markowitz, J.A.: Using Speech Recognition. Prentice Hall PTR, Upper Saddle River (1996)
5. Porta, A., Deru, M., Bergweiler, S., Herzog, G., Poller, P.: Building multimodal dialog user interfaces in the context of the internet of services. In: Towards the Internet of Services: The THESEUS Research Program. Part of the series Cognitive Technologies. Springer International Publishing, Heidelberg, pp. 145–162 (2014)
6. Potapov, V.: Speech rhythmic patterns of the Slavic languages. In: Ronzhin, A., Potapova, R., Delic, V. (eds.) SPECOM 2014. LNCS, vol. 8773, pp. 425–434. Springer, Heidelberg (2014)
7. Potapova, R.: Priority trends of the present day applied linguistics. J. Convers. Mach. Build. Russ., pp. 3–4 (2004) (in Russian)
8. Potapova, R.K.: On Natural Language Processing Technology on the Domain of Science & Industry. Russian Academy of Sciences, Ozyorsk (1992). (in Russian)
9. Potapova, R.K.: Speech Driving of Robots, 2nd edn. KomKniga, Moscow (2005). Revised and corrected (in Russian)
10. Potapova, R.K.: Speech: Communication, Information, Cybernetics, 4th edn. Librokom, Moscow (2010). (in Russian)

11. Rigoll, G.: Multimodal human-robot interaction from the perspective of a speech scientist. In: Ronzhin, A., Potapova, R., Fakotakis, N. (eds.) SPECOM 2015. LNCS, vol. 9319, pp. 3–10. Springer, Heidelberg (2015)
12. Saraclar, M., Dikici, E., Arisoy, E.: A decade of discriminative language modeling for automatic speech recognition. In: Ronzhin, A., Potapova, R., Fakotakis, N. (eds.) SPECOM 2015. LNCS, vol. 9319, pp. 11–22. Springer, Heidelberg (2015)
13. Saveliev, A., Basov, O., Ronzhin, A., Ronzhin, A.: Algorithms for low bit-rate coding with adaptation to statistical characteristics of speech signal. In: Ronzhin, A., Potapova, R., Fakotakis, N. (eds.) SPECOM 2015. LNCS, vol. 9319, pp. 65–72. Springer, Heidelberg (2015)
14. Schmandt, C.: Voice Communication with Machines. Conversational Systems. Van Nostrand Reinhold, New York (1994)
15. Wechsung, I.: What are multimodal systems? Why do they need evaluation? Theoretical background. An evaluation framework for multimodal interaction. Part of the series T-Labs Series in Telecommunication Services, pp. 7–22 (2014)

The Humanoid Robot Assistant for a Preschool Children

Alina Zimina, Dmitry Rimer, Evgenia Sokolova, Olga Shandarova,
and Evgeny Shandarov[✉]

Laboratory of Robotics and Artificial Intelligence,
Tomsk State University of Control Systems and Radioelectronics (TUSUR),
Tomsk, Russian Federation
evgenyshandarov@gmail.com

Abstract. The prototype software for humanoid robot assistant for a preschool children was developed. Aldebaran NAO robot was used for this development. This paper describes the idea, main concepts and applications for robot.

Keywords: Robot–human interaction · Aldebaran robotics NAO · Robot for preschool children

1 Introduction

Social Robotics — a new direction in science, psychology and technology, including robotic systems in social interactions. Social robots - is a promising direction, both in terms of research and business.

Scope of social robots is extensive, it includes robotic personal assistant, companion robots for the elderly, the robots to work with children, etc. It looks promising that anthropomorphic robotic platform Pepper designed specifically for use in home.

Robot-Child Interaction could be based on human-centered interaction [1]. An example of this can serve a number of projects designed to investigate the interaction of children with robots. For example, the project «Aurora» is exploring the use of robots for the purpose of therapy and education of autistic children [1], and the project «ALIZ-E» is dedicated to the study of long-term interaction of the robot with children in the clinical examination [2]. The very promising approach in human-computer interaction is an assistive Bi-modal interfaces where speech recognition and computer vision are used cooperatively [3].

Robots are created to solved special tasks. However, in recent years it has been a trend, when developers create so-called universal robotic platform. Functionality of such robots depends on loaded set of applications. It is opening opportunities for independent developers and making household robots more attractive from a consumer perspective. Applications can be installed to robot, through the online stores, like Google Play Store and Apple AppStore.

© Springer International Publishing Switzerland 2016
A. Ronzhin et al. (Eds.): ICR 2016, LNAI 9812, pp. 219–224, 2016.
DOI: 10.1007/978-3-319-43955-6_26

2   Implementation and Experiments

2.1   "The Babysitter Robot" Project

The "Babysitter Robot" project has been developing in robotics and artificial intelligence lab TUSUR since 2012. The aim of the project is to create hardware and software that implements the functions of an assistant kindergarten teacher. The idea is that kinder-gartener can not provide enough attention to all children in the group during the day. "Babysitter Robot" should help in this case, by interaction with children at least for part of the day. The goal of the project is the creation and implementation of robotic software and hardware in social work with children of preschool age.

The architecture includes anthropomorphic robotic platform with advanced commu-nication systems, a set of applications implementing the various scenarios of human-robot interaction and program-manager, which providing launch of the desired application.

2.2   Robot Hardware

The project implementation uses a humanoid robot, which should provide the necessary level of emotional connection to between child and robot. The selected platform is ambiguous, because of the large different configurations of robots. Based on a set of specifications for the task the robot Aldebaran Robotics NAO has been used. It has: 25° of freedom; Sensors: 2 cameras, a 4 microphones, 9 tactile sensors and 8 pressure sensors; communication: voice synthesizer, LED indicators and 2 speakers; connection: WiFi and Ethernet. Choreographe as environment for software development.

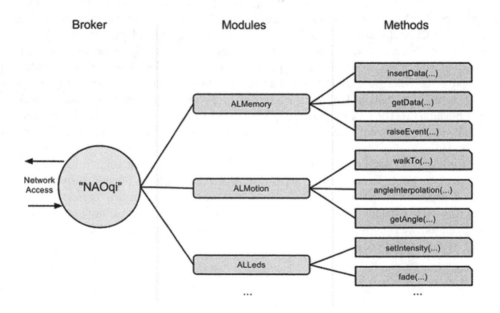

Fig. 1. NAOqi framework system libraries

NAOqi library provides tools for developers to implement character recognition and image, voice control and the possibility of synthesis (Fig. 1). NAOqi main modules are: Core - includes memory management techniques, a compound with a computer, etc.; Motion - includes methods to controlling robot movement; Audio - module to work with sound; Vision - Imaging module; Sensors - sensor processing.

2.3 The "Babysitter Robot" Applications

As was shown in our previous researches [3, 4], for efficient human-robot interaction is important to write a perfect scenario. From 2012 we have developed several dozen child-robot interaction scenarios, some of them after testing on focus groups have been approved to include into project.

The app **"Robot Control"** allows the child to learn the basics how to control the robot and overcome the psychological barrier in the interaction with the machine. Child control robot by toy, which has solid contrasting color, such as red cube. According to the scenario, the robot has to maintain a certain distance to the toy, about 1 m. If red cube moves from the robot, robot moves to target, if closer - steps back. In this mini-game the child has task simply move robot on any trajectory. NAOqi Vision library functions was used to find the contrast object, based on the original image binarization method. Direction detected according to the location in the center of mass in the frame of monochrome pixels and distance by the ratio of the volume of these pixels to the total number of them in the frame. The tests were conducted at various exhibitions (Fig. 2).

Fig. 2. Children control robot walking (Color figure online)

The **"Excersises"** application is designed to provide complex of physical exercises. The scenario of app is, the robot greets the children and offers to repeat the exercise from him.

The application includes following exercises: turns heads in different directions; warm hands, elbows, shoulder joints; warm legs, waist; last exercise – "martlet" (Fig. 3). Robot-child interaction based on robot voice module. Movements of robot were created in Choreographe software. This application provides the most favorable response of both children and adults during the demonstration at presentations and exhibitions. Children of different ages enjoy doing the exercises together with the robot.

Fig. 3. Children are doing excersises with robot

The application "**Action Poems**" based on the pedagogical practices method when working with children of a half to 5 years. Kids like to illustrate the poem by actions, it expands their vocabulary, improve coordination, creative thinking and imagination, trains attention and memory. We have used a simple rhymes for application, such as "The deer big house…". Independent animations were created for each stanza and transitions between them. An important parameter is the time to synchronize each stanza and animation. Since the time required to perform the action is bigger than time needed for "reading", the synchronization was performed on execution of actions. Tests were carried out in the laboratory and at the "RostUp-2014" exhibition. The demonstration at the exhibition was held in front of focus groups - children aged 3 to 7 years. Tests have shown results such as acceptance by the target audience, and was positive. Most of the children diligently repeated robot actions. The robot animations in general, are quite visible and easily understood by children. Motion Modules and Audio NAOqi library were used.

"**Traffic Officer**" The application helps children learn the traffic signs in a game form. The scenario includes the following steps: robot greeting; offer to learn and repeat traffic signs; demonstration of the cards to children, then to the robot; robot scan cards;

read out the text corresponding to the image on the card; if the card is unfamiliar or is shown incorrectly, the robot says he does not know what is illustrated on card; at the end of the lesson, the robot thanks children for their participation and bids farewell.

To implement this idea used features such as a robot voice synthesis text, vision system and image recognition (NAOqi modules Audio, Vision, Motion). Image recognition performed based on the comparison with the reference image stored in the memory of the robot. Creation of reference images produced manually. The tests were carried out in the laboratory, at exhibitions and presentations. The app demonstrated high accuracy and speed of character recognition.

Application "**The game of emotions**" shows children the ability of robot to demonstrate a variety of emotions. Scenario: children show pictures with images of emotion; they determine what is this emotion; then show the picture to the robot; robot recognizes picture, and performs typical actions for that emotion. The robot recognizes and demonstrates the following emotions: happiness, surprise, anger, sadness, fear. NAOqi Vision and Motion modules was used (Fig. 4).

Fig. 4. The game of emotions

Application "**Arithmetic**" supposed to help children learn simple mathematical functions in a playful way. This program implements a simple mathematical operation (addition, subtraction, multiplication, division). The scenario is as follows: children show cards with numbers and mathematical signs; They say the answer; then demonstrate the robot card, which gives the answer (Fig. 5).

The app manager allows you to choose an application to execute by oral and tactile interaction with robot. NAOqi Core, Audio and Sensor modules was used. Execution of any application can be stopped and interrupted by pressing tactile sensor.

Fig. 5. "Arithmetic" application

3 Conclusion

As a result of the work we have developed a prototype of software that implements the functions of the robot-assistant preschool teachers, developed scripts interaction child-robot and robot-children. The developed applications can be used in the advanced systems based on commercial robotic platforms for use in home, such as Pepper robot by Aldebaran Robotics and Softbank.

References

1. Dautenhahn, K.: Socially intelligent robots: dimensions of human–robot interaction. Philos. Trans. R. Soc. Lond. B Biol. Sci. **362**(1480), 679–704 (2007)
2. Belpaeme, T., Baxter, P., Wood, R.: Multimodal child-robot interaction: building social bonds. J. Hum.-Robot Interact. **12**, 33–53 (2012)
3. Karpov, A., Ronzhin, A., Kipyatkova, I.: An assistive bi-modal user interface integrating multi-channel speech recognition and computer vision. In: Jacko, J.A. (ed.) Human-Computer Interaction, Part II, HCII 2011. LNCS, vol. 6762, pp. 454–463. Springer, Heidelberg (2011)
4. Shandarov, E.S., Zimina, A.N., Ermakova, P.S.: The behavior robot-assistant analysis within the framework development scenarios child-robot interaction. Humanitarian Inform. **8**, 52–64 (2014). (In Russian)
5. Zimina, A.N., Ermakova, P.S., Shandarov, E.S.: Robot-assistant behavior analysis for robot-child interaction. In: Innovations in Information and Communication Science and Technology, Third Postgraduate Consortium International Workshop (Tomsk), pp. 119–127 (2013). (In Russian)

Voice Dialogue with a Collaborative Robot Driven by Multimodal Semantics

Alexander Kharlamov[1]([✉]) and Konstantin Ermishin[2]

[1] Institute of Higher Nervous Activity and Neurophysiology of RAS, Moscow,
Russian Federation
kharlamov@analyst.ru
[2] Bauman Moscow State Technical University, Moscow, Russian Federation
konstantin.ermishin@gmail.com

Abstract. The paper describes the control system of an autonomous
mobile service robot using a human-machine dialogue to control the robot
in a way natural for human with the help of syntactic structures that
define the sequence of certain actions. The interaction with the robot
by means of natural language in the formation of target designations is
provided by the multimodal representation of the working area in the
form of a hybrid map of the surrounding space with highlited work areas
and locations of key objects complemented by a set of semantic relations that characterize features of the robot operation and impact of the
environmental changes on its operation. The article discusses the control
system of the mobile service robo, performing transport and logistics
operations in a public place in direct contact with lots of people around.
The proposed approach to the dialog control of the robot based on the
multimodal representation of the robot operation area not only simplifies
the process of determining the sequence of operations, but also enhances
the level of safety for other people, objects and the robot itself in the
process of fulfilling its service tasks.

Keywords: Collaborative robots · Autonomous navigation of a mobile
robot · Path searching · Hybrid terrain map · Safety of movements of
a mobile robot · Multimodal model of the world · Semantic network ·
Voice dialogue

1 Introduction

Currently, the use of service robotics becomes quite extensive. Unlike industrial robots, usually operating autonomously without human involvement, service robots are most often used in solutions of collaborative tasks performed
together with a human partner. Working together with a human partner does
not imply reducing the degree of autonomy of the service robot, and rather
imposes a set of specific safety requirements to the working process and simpli-
fying methods of human-machine interaction.

© Springer International Publishing Switzerland 2016
A. Ronzhin et al. (Eds.): ICR 2016, LNAI 9812, pp. 225–233, 2016.
DOI: 10.1007/978-3-319-43955-6_27

Therefore, the robot control system must not only ensure its intended functions, but also respond adaptively to changing environmental conditions and commands received from the human partner. In this regard, the control system must be able to simultaneously compare and analyze a large amount of various information service operation parameters, environmental changes and their impact on the robot operation, as well as assignments received from the human partner. The degree of adaptability of the robot depends on the ability of the robot control system to comprehensively assess the situation and to form the most current control command.

The complexity of the information analysis lies in a variety of used data describing a task. For example, to set a path of the mobile robot a geometric description of the working area is used presented in the form of an array of coordinates describing areas of allowed and restricted movements. The objects which interact with the robot are described by coordinates of the working area and with the help of parameters characterizing the object properties and the number of possible operations. In turn, the human partners commands are specified as a set of instructions in natural language describing the action, the place of its implementation and the object with which it is necessary to interact. As a rule, this kind of data is rarely associated with each other, which reduces the efficiency of situation evaluation and hampers the collaborative human-robot interaction.

Currently for resolving of this contradiction the multimodal semantic representation of environment is using, which includes composition of the navigation information (geometric description of the work area - allowed and forbidden for traffic areas, descriptions of objects with which communicates robot), as well as the visual representation of objects as parameters of their images (images or video of certain views of environment) [1]. These views are complemented by semantic and topological network which unites them and adds the names of objects presented on the navigation plane. Recognition of visual images of objects improves the task of control process.

Using multimodal semantics, which includes the plan of rooms and a semantic network that characterizes objects relations and visual images of objects, enables effective control of the robot accompanied by visual control (inspection) of the task. Control efficiency increases significantly when a dialogue is introduced in the control circuit. Within the framework of human-robot dialogue there are issues related with recognizing of the meaning of the sentences addressed to robot by natural language [2], which solves by using of special dimensional language for description processes and effective understanding task [3].

However, the dialogue is conducted by means of traditional computing: the robot commands are given from the keyboard, and the responses for human-companion are given on the monitor screen. Of course, using voice subsystem is improving conditions dialog interaction between the robot and human-companion, because the robot transmits the answers automatically by speech synthesis, and recognizes human speech commands. The speech input and output allows to avoid the close connection between human-companion and the robot and simplifies the process of communication with him [4]. There is an opportunity to communicate with the robot at a distance by means of, for example, a headset with Bluetooth.

The paper presents a control system of an autonomous mobile service robot exercising collaborative interaction with a human partner by speech dialogues on a subset of natural language and performing transport and logistics operations in a public catering facility in conditions of probable contact with visitors or surrounding objects. A special feature of the control system is the robot's capability to be engaged in an intelligent dialogue in preparation and execution of tasks. An intellectual dialogue is possible with the use of a multimodal model of the robot world, including the premises plan, a semantic network describing the thing world of the robot in the form of interrelations of its objects and supplemented by visual information about objects surrounding the robot. The feature of such a world model is combination of various modal representations of the model by overlaying the semantic network on the premises plan, which makes it convenient to visualize the world model, which in turn is ergonomic in terms of storage of this information in the human partner's memory. Naming the nodes of the semantic web with not only lexical labels, but also using their speech equivalents simplifies the use of semantic web for implementation in control system of collaborative robot.

2   Mobile Service Robot Control System

The paper deals with the process of improving the control system of a mobile service robot OBYS designed for transporting goods from the main hall of a restaurant to technical premises.

A mobile robot OBYS (Fig. 1) was used which is a differential chassis equipped with a laser scanning range finder Hokuyo UTM-30LX, a control system based on a single board computer ODROID-XU4 and hardware-software system ST-Robotics for autonomous navigation inside buildings.

The robot operation is performed in cooperation with a human partner serving the restaurant hall, who controls the robot by voice commands in natural language and loads the robot with trays with dishes transported to technical premises.

In addition to interacting with a human partner, in the course of work the robot interacts with other people (visitors of the restaurant and service personnel) detecting them on its way and planning the route in order to ensure safe movements. Since the robot moves in space-limited environment due to humans constantly moving in close proximity, the robot control system performs some situation evaluation during the formation of interaction with it.

The mobile service robot control system is a two-tier architecture consisting of a base unit and an expansion unit (Fig. 2). The base unit of the control system is responsible for controlling the mobile robot's movements; it collects and analyzes sensory data, generates a map of the working area; localizes the robot and surrounding objects; plans the route and controls the robot's movements.

The expansion unit of the control system provides a mechanism for collaborate human-robot interaction by means of a natural language dialog unit which

Fig. 1. Mobile service robot OBYS

is used by the human partner to form a sequence of working tasks for the robot. In the formation of a natural language task its description is assocoated with a topographical description of the work area indicating the geometric coordinates on the premises plan, or any objects and actions with them. For example, the task to move to the technical premises associates the geometric coordinates of the target point with the topographical map of the premises plan. On the other hand, the task to stay in the restaurant hall to load trays associates the topographical plan of the premises with many potential parking spaces, as well as the robots current position and a comprehensive evaluation of the work area, taking into account the position of surrounding objects, movements of people and safe movements of the robot.

2.1 Hybrid Map of the Surrounding Space

Safe movement of the mobile service robot and effective interaction with a human partner is possible through the use of multimodal representation of the surrounding space in the form of a multi-layer map of the robot's working area (Fig. 3).

The surrounding area is represented as a set of dat from the topographic plan, safe areas and restricted areas, and the local map based on data from the robot-borne sensor systems [5]. The topographic plan is a geometric plan of the premises with areas marked as restricted areas and areas of work operations. Integration of information from a variety of local maps obtained from data of the robot-borne sensor systems, is used to determine the environmental conditions. For example, a map obtained from the laser scanning range finder is used to localize the robot on the topographic plan, and a map obtained from

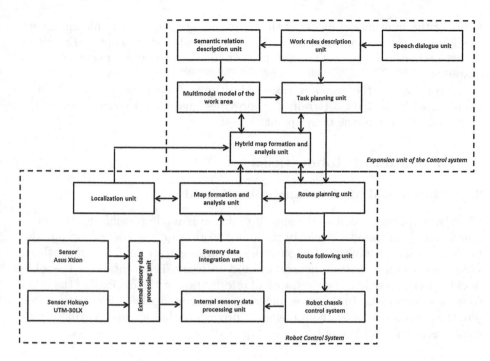

Fig. 2. Architecture of the mobile service robot control system

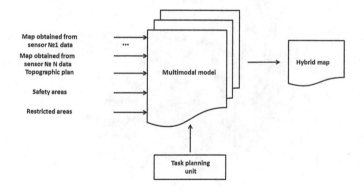

Fig. 3. Multimodal representation of the working area

a set of sonars located on the perimeter of the robot, enables determination of objects around the robot during its movement [6]. As a result, the evaluation of changes in the working area status together with the assigned task allows to create a sequence of safe maneuvers of the robot to ensure its movement in rapidly changing space-limited conditions.

Robot route planning is provided with the use of a hybrid map, which is an artificially created terrain map obtained by aggregation of individual map layers of the multimodal model of the surrounding space. Route planning with the hybrid terrain map provides the capability to calculate the optimal route for the current situation, as well as automatic and safe avoidance of collisions with obstacles.

In contrast to the use of various algorithms of obstacle avoidance, route planning with the hybrid terrain map provides guaranteed convergence of the movement process to the target point.

3   Formation of Robot Tasks

3.1   Multimodal Model of the Robot World

Effective implementation of the speech dialogue is made possible by the introduction of a multimodal model of the robot world into the control circuit; this model is based on multimodal semantics. Using separate modules of synthetic vision, speech input-output, networking and ontological conceptions of the robot world does not provide efficient use of all information at the disposal of the robot to solve service tasks. In this case, inclusion of the speech dialogue with the robot into the control circuit is a non-intelligent procedure that enables, virtually, only solutions to the problem of speech control.

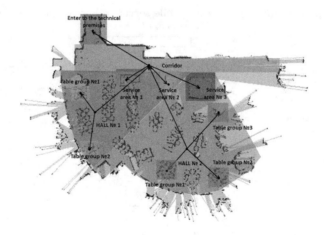

Fig. 4. Semantic network projected on the premises plan

Combining all representations of these modalities into a single multimodal semantic network fundamentally changes the situation: individual components of the representation complement each other allowing to implement an intelligent dialogue. The robot world model includes: (1) a semantic network [7], the lower-level vertices of which are objects of the premises in which the robot operates, associated with top-level vertices corresponding to these premises; (2) video images of objects (photo pictures or special graphic labels uniquely marking an object) located in the premises associated with lower-level vertices; (3) speech images of denotations for these objects that are used for speech description of objects.

The semantic network describing the robot world model is projected onto the premises plan, which includes coordinates of the objects depicted on it (see Fig. 4). This representation convenient for its visualization to the human partner is natural and ergonomic for its memorizing and storing in the human partners memory.

3.2   The Robot Movements Planning

The implementation of multi-modal model of the environment allows us to consider the working area of the robot is not like an ordinary plan of the premises, but as well as a collection of different working areas, endowed with a number of custom features. In this work, working areas are endowed with features characterizing features of the robot movement, for example - speed limit zones in each of the rooms of the restaurant, or areas showing the dangers of movement of the robot due to a possible clash with the restaurant staff or other robots. The ability of the robot to effectively respond changes in the environment during it's movement enables secure collaborative interaction with the staff.

The multimodal model in addition to providing of the floor plan as a set of working areas include more semantic links between the different zones. As a result, it becomes possible to describe not only the conditions for movement of the robot in a certain area, but also the conditions of the transition between the zones. In particular, the semantic connections determine the direction of movement of the robot, allowing you to advance consider the influence of the environment in all the way of the robot motion.

The following shows an example of a movement from corridor to the technical premises:

IF ("Start position") $\in$ Corridor) $\cup$ ("Finish position" $\in$ Technical premises) $\rightarrow$ Limitation of movement to the right

For example, when the robot moving along a corridor to the technical room during his way may be staff coming from the hall to the left. In this case, the robot usually performs a maneuver to the right, but there is another way out of the hall on the right side of the hallway, as a result of the maneuver can lead to a collision with the staff or visitors of the restaurant.

IF ("Start position" $\in$ Corridor) $\cup$ ("Finish position" $\in$ Table group No 2) $\rightarrow$ (Speed reduce in Hall No 1) $\cup$ (reducing the size of the safety zone) $\cup$ (prohibition of maneuvers)

When the robot moving from the corridor into the hall No 1 it overcomes several narrow passages between the tables, among which visitors and restaurant staff can move. To improve the safety of the robot required reduction velocity, reducing the security zone around the robot and limitation of maneuvers.

Zones of restriction maneuvers and safety zones are fragments of hybrid map, each of which is added to the map of the floor plan terrain and local maps of different sensory subsystems. As a result of aggregation of data is formed hybrid map that takes into account possible changes in the environment and safety strategy robot.

3.3   Dialogue with the Robot

Since the dialogue concerns mainly issues of movements, dialogue patterns are limited to movements from the starting point to the target point of the premises. Both points are points of the premises marked with their coordinates. The route planning unit uses premises plans of the hybrid map creates the most effective ways of moving for the mobile service robot.

Control commands for the robot are formed in the task planning unit using the information of the current demand for service tasks, target designations received from the human partner and the state of the environment. Control commands in the whole are reduced to movements from location points of the restaurant halls, through the corridor, to the technical area, and vice versa, and to movements between location points in the restaurant halls. That is, tasks are relatively simple. These tasks are set in a number of movement programs, the specific parameters of which may change in the process of dialogue.

The dialog template in this case is extremely simple: (Rob), take (trays), Shall I come back?, Yes, (to the) table number 6 (in) hall number 1, and other similar commands, questions and answers. The dialogue allows entering coordinates of the target point, or both initial and target points. In the process of movement conflicts may occur that are resolved by addressing a question to the human partner. This can be done remotely, as a Bluetooth headset is used. These collisions may relate to both unforeseen obstacles on the robot route and other unforeseen situations. In extreme cases, the robot may address to his human partner who then takes the lead.

3.4   Speech Input-Output

To implement a speech dialogue with the robot, a system of command speech recognition is used [4]. A fixed headset microphone is used that improves recognition quality and also allows display robots speech messages on the phone of the headset. Speech output is just playing responses recorded and downloaded in the memory of the system.

Recognition of single commands is easy and convenient for use technology in this case, also this is not very burdensome for the human-companion in terms of training, because the dictionary, which describes all possible operations is not large. It requires pronouncing the words of the dictionary of control system before the start of each regular session of communication with the robot. On the other hand, there is an adjustment to speaker like as an optional feature of protecting against unauthorized communication with the robot.

4   Conclusion

The paper describes an integrated mobile service robot including modules of navigation, control, representation of the world models, and dialogue implementation in its architecture. Using all available sensory data presented in an integrated form as a multimodal model of the world allows implementation of effective intelligent speech dialogue of the human partner with the robot during execution of tasks both in the process of task formulation and in the process of its implementation, especially in emergency situations, which greatly facilitates the work of the human partner and the robot. If a speech intelligent dialogue subsystem is available, even an untrained person can work with the robot, which plays a significant role in terms of the profitability of service robots.

Acknowledgement. The research was performed under the project Study of the mechanism of associative links in human verbal and cogitative activity using the method of neural network modeling in the analysis of textual information (financially supported by the Russian Foundation for Basic Research, grant N 14-06-00363).

References

1. Lim, G.H., Suh, I.H., Suh, H.: Ontology-based unified robot knowledge for service robots in indoor environments. Trans. IEEE Syst. Man Cybern. Part A Syst. Hum. **41**(3), 492–509 (2011)
2. Heriberto, C., Nina, D., Kai-Florian, R., Thora, T., John, B.: A dialogue system for indoor wayfinding using text-based natural language. Int. J. Comput. Linguist. Appl. **1**(1–2), 285–304 (2010)
3. Skubic, M., Perzanowski, D., Schultz, A., Adams, W.: Using spatial language in a human-robot dialog. In: IEEE International Conference on Robotics and Automation, vol. 4, pp. 4143–4148 (2002)
4. Zhonin, A., Kargin, D., Sergiyevskiy, N., Kharlamov, A.: Model mira robota na osnove sensoriki blokov navigatsii, tekhnicheskogo zreniya i rechevogo dialoga. Trudy mezhdunarodnoy molodezhnoy konferentsii Informatsionnyye sistemy i tekhnologii, pp. 68–69 (2012)
5. Ermishin, K., Vorotnikov, S.: Intellektualnaya sistema upravleniya servisnym mobilnym robotom. Politekhnika-servis, Ekstremalnaya robototekhnika. Trudy vserossiyskoy nauchno-tekhnicheskoy konferentsii, Sankt-Peterburg (2012)
6. Ermishin, K., Vorotnikov, S.: Multiagentnaya sensornaya sistema servisnogo mobilnogo robota. Vestnik MGTU im. N. E. Baumana. Ser. Priborostroyeniye **6**, 50–59 (2012)
7. Kharlamov, A.: Neyrosetevaya tekhnologiya predstavleniya i obrabotki informatsii (estestvennoye predstavleniye znaniy). Radiotekhnika (2006)

Volumetric Display Testing Unit
for Visualization and Dispatching Applications

Alexander Bolshakov[1(✉)], Arthur Sgibnev[2], Tamara Chistyakova[3],
Viktor Glazkov[2], and Dmitry Lachugin[2]

[1] Peter the Great St. Petersburg Polytechnic University, Saint Petersburg, Russia
aabolshakov57@gmail.com
[2] Yuri Gagarin State Technical University of Saratov, Saratov, Russia
aasgibnev@gmail.com, glazkovvp@gmail.com, lachugindm@mail.ru
[3] St. Petersburg State Technological Institute, Technical University,
Saint Petersburg, Russia
chistb@mail.ru

Abstract. We present software and hardware solutions for volumetric display testing unit with projector array. We designed workstation based on sixteen single-board computers Raspberry Pi 2 Model B, which connect our testing unit to the user computer. Also there is provided the interface to load and control three-dimensional models in *.3ds* format through user computer. Our experimental system provides sixteen images with 800×600 resolution for forming volume in $160°$ visibility in horizontal direction. Our developing project will be useful for application in simulators and for solving CAD tasks in mechatronic systems. Further, it is planned to create a volumetric display with all-round visibility of dynamic full-color images with real-time visualization for dispatching applications.

Keywords: 3D-visualisation · Multiscreen system · Projector array · Volumetric display · Dispatching · Mechatronic systems · Raspberry

1 Introduction

There are nine scientific centers and thirty largest corporations had developed autostereoscopic volumetric and 3D displays from 2009 to 2014 with different level of success. However, developed models are not widespread due to poor image quality and high product prices. Nevertheless, according to studies, there is a significant increase of an efficiency in various activities when applying 3D-technologies. According to the analysis agency Gartner section of volumetric (holographic) displays got into the promising technologies that are in its infancy. It will be a significant growth of volumetric displays market. It is waiting for a replacement of 3D-displays samples for a comparable price and with characteristics, which will exceed their possibilities [1].

Special glasses or virtual reality helmets are applied for 3D visualization. They use relatively expensive and complex technical solutions for 3D

© Springer International Publishing Switzerland 2016
A. Ronzhin et al. (Eds.): ICR 2016, LNAI 9812, pp. 234–242, 2016.
DOI: 10.1007/978-3-319-43955-6_28

demonstration without stereo glasses for mass audience. Among this are planar autostereoscopic displays with a limited number of viewpoint and 'air' displays, reproduced 'hanging in the air' images.

In this regard, there exists the problem of quality improvement of color and resolution, also there problem of limited size of formed three-dimensional image and reducing the cost of the product. Additionally there are needs of increasing of informativeness of content images and reducing the proportion of computer processing requirements for the data channels to the hardware and software characteristics of the product.

This project is relevant to monitoring and control systems market because visual aspect is essential component for operational activities and monitoring of fast moving objects. Three-dimensional technology displays depth and/or height of the object and transmits parts which inaccessible for two-dimensional technology. Promising areas include the decision support in the design and management of various technological objects or technical and organizational systems, based on the volume dynamic and full-colored imaging. In addition, the produced product will be useful for application in simulators and for solving CAD tasks in mechatronic and interactive systems with three-dimensional images.

Our project is associated with formation of the relevant hypotheses and solution a number of interrelated problems. This article describes the current state of the project.

2 Setting of the Problem

Currently, volumetric displays have not solved the following problems: (1) the demonstration of full-color three-dimensional objects in good quality with all-round visibility for mass audience, which do not use stereo glasses; (2) realization of real-time volumetric broadcasts with network connection. Existing systems have information flow required to reproduce the 3D-images, exceed the capacity of existing communication channels.

These problems are complicated and their solution needs step-by-step implementation. In the first stage, we conduct a number of experiments (physical and computing), which confirmed the validity of basic ideas. Next step (current stage) is associated with the validation of hypotheses based on design of testing unit for volume visualization.

To implement the testing unit it is necessary to form 3D images from pre-arranged content and display it with 800×600 resolution on the array of 16 projectors. Projectors form three-dimensional image seen in the range of $160°$ horizontally (in the future we are going to increase the number of video outputs and the image resolution). This task is decomposed into a series of interconnected subproblems. For this purpose, in particular, we need to offer and design solutions for hardware and software implementation of the testing unit. Scientific and technical originality of proposed solutions and the corresponding scheme of testing unit for volume visualization are briefly described in Sects. 3 and 4 respectively.

3   Scientific and Technical Originality of the Proposed Solutions

We have proposed the volumetric display [2,3], which is based on a patented technology combining optical image processing with computer. The following principles can be underlined in volume image forming technology: (a) combined application of optomechanical and computer image processing systems; (b) multiple image sources, each of which render dynamic images of the object of corresponding azimuthal position; (c) optical properties of testing unit allows to obtain a large amount of non-recurring intermediate images, uniformly filling all azimuthal observation space. And it is possible with a small number of reference images; (d) anisotropic screen increases the observed brightness of the image and provides the autostereoscopic effect.

Scientific and technical novelty of the proposed solution and the key differences between analogues are as follows: (1) set of output devices that is different from existing solutions which flow of data is restricting and it could affect the image quality and causing the use of devices with maximum performance. This allows us to use hardware with undemanding requirements; (2) intermediate images continuously and with equally density fill all azimuthal space of object and to provide continuity of pattern in azimuth. That allow us to use a limited number of reference images of the object (not more than 72); (3) the total information flow does not exceed the real throughput of standard computers ($\sim 100\,\text{Gb/s}$; with restriction of number of reference images to 80), which distinguishes it from functioning analogue, in which number of reference images is not less than 180 and the flow of them has to restrict by the decline of resolution and color of images. It is allows in our case to provide the transfer of full color reference images in FHD; (4) image processing, based on combination of two reference images and optical system allows to 'unload' the output device for high-quality images with large size and high resolution.

As can be seen from the above, key benefits of proposed solution, based on a combination of computer and optical image treatments, in comparison with the analogues, introduced to the market or under the development, are as follows: (1) 24 bits color depth is equal or higher in comparison with analogues (8–24 bit); (2) reduced in two–six times data flow for normal work; (3) low requirements for GPU performance and system output; (4) it is possible to use inexpensive and accessible components.

4   Testing Unit Description

For the validation of our hypotheses, we carried out the work on the developing of three-dimensional visualization testing unit. Around the axis 1, as shown in Fig. 1, rotates opaque cylinder 2 at the rate of 1200–1500 rpm. The cylinder has through holes 3, closed by transparent material. Inside the cylinder is sets a translucent lenticular screen 4, which displays the volume image.

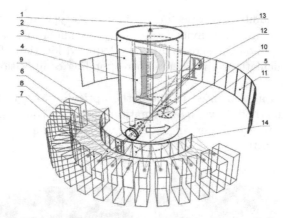

Fig. 1. Testing unit structure diagram

On the front side of the testing unit is transparent multi-screen panel 5, designed as a direct part of a multifaceted prism and coated with rear projection film, which shows the primary image 6. The image forms on all the faces of multi-screen panel with an array of DLP-projectors 7. Beams of light 8 focus on the rear projection film layer.

Two parts of the optical system 9 and 10 send light beam, which crosses the axis of the testing unit and reaches the surface of the polygon mirror 11, composed of a flat mirror surfaces. After reflection from the polygon mirror, the light beam reaches the screen 4 and forms the resulting image. Electric motor 14 rotates cylinder with bearings 13.

This testing unit is the prototype of a circular volumetric display for 3D visualization.

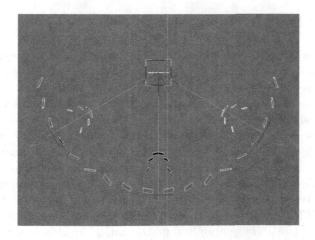

Fig. 2. Scene in 3Ds Max, top view

To reduce the number of projectors, in first prototype unit we will use three projectors, wich images are divided into 18 parts and projected through the system of 36 mirrors on the screen (Fig. 2). Required location of the mirrors is determined by calculations and then a corresponding composition was made in 3ds Max program.

This scheme is implemented to check the optical system of the display. In addition to the method of volume formation with superposition of images of object, recorded from different angles, also we tested the algorithm of sliced image forming (Fig. 3).

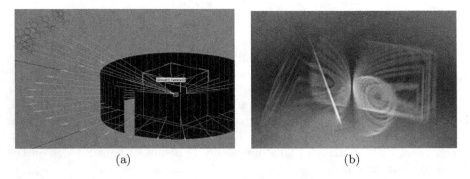

(a)   (b)

Fig. 3. Experiment of image formation from slices: (a) scene in 3Ds max; (b) visualization of scene with primitives from slices

Above-described experiments and modeling focused on analysis of different ways of volume formation and construction optimization of the optical system and the testing unit design generally.

5   Testing Unit Architecture

The project realization is associated with the formation of relevant hypotheses and solution of interrelated problems. One of the current tasks, which is devoted to this article, is the image output to multiple devices, such as monitors or projectors. This scheme can be used in the game industry to expand the virtual space, for video wall visualization and so on. These solutions are used in 3D visualization tasks. It is also reasonable to create testing units for hypothesis verification of characteristic determination of processes, that occur in the formation of three-dimensional image. In addition, we need to consider the requirements for speed and flexibility of the respective architectures of information-computer systems, during the development of our testing unit. Moreover, we need to determine the methods of image formation and control for content, created for autostereoscopic displays [4–6] and use approved mathematical apparatus for image model transformation [7].

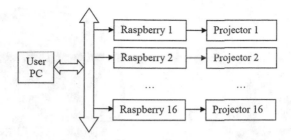

Fig. 4. Scheme of output images using single-board computers

As previously noted (Sect. 2) for volumetric testing unit implementation is required to form a three-dimensional image by prearranged content output with defined resolution to required projector array. Thus formed three-dimensional image seen in a range of 160° horizontally. Our solution for software-hardware part of volumetric visualization testing unit are listed as follows.

After evaluating of various ways to form and content control, we decided to use the client-server connection method (Fig. 4), because it has sufficient flexibility, scalability and acceptable for this task data transfer rate.

We use single-board computers to solve the problem of high cost. For the experiments and evaluation of performance of the proposed scheme, we select a single board computer Raspberry Pi 2 Model B, which has increased performance, in comparison with older models. At an early date, we are going to create a workstation (Fig. 5). Currently, performance capability tested with two single-board computers.

Our system implements three operation modes: static image demonstration, video playback and *.3ds* model loading and control. Let us take a closer look to the last mode.

6   Testing Unit Software Development

The program on user computer opens the prepared 3D model in *.3ds* format, which user can move, zoom and rotate by mouse. Simultaneously with the opening of a model on the user's computer, this model is also loaded into the workstation. Depending on the projector location, the initial position of the model in the workstation moves to a fixed angle of rotation. Any changes in model state generated event (changing of mouse coordinates, pressing buttons, etc.)

This event, in turn, alters characteristics of the respective model and changes a model matrix, which is transmitted over the network to workstation.

User computer performs the following functions: reading of a 3D model; interworking interface; scene mount; manipulation with the model in the scene; network transmission of the matrices in workstation.

The workstation performs following functions: receiving data from the user computer for changing the model state; angular 'binding' of the virtual camera

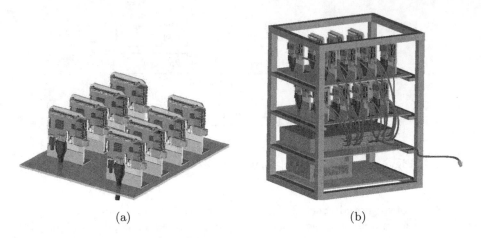

(a)   (b)

Fig. 5. Workstation CAD model with Raspberry boards: (a) boards arrangement on the same shelf; (b) general view

in accordance with their position; on-screen display of graphic object; synchronization of image outputs.

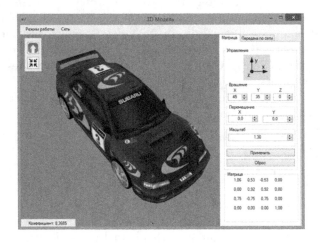

Fig. 6. The application window with an open textured model

Software application for reading and control 3D models on the user computer is implemented using SharpGL library - it is OpenGL for .NET applications. Workstation application is developed for OS Raspbian in cross-development environment CodeTyphon. To read the three-dimensional object is selected 3DS format, as it is quite common and well documented. 3DS file consists of data blocks (chunks), which has a standard structure within the format.

Chunks form a tree structure; one block may contain several other sub-blocks. Each unit has the following format: block header – 2 bytes, the size – 4 bytes, data – n bytes and sub-blocks – m bytes.

Currently we implement the reading of models with textures in different formats (Fig. 6). In the future we plan to add the ability to play animation, assembly/disassembly of models and function of hiding/showing model details.

Content management from the user's computer is based on the resulting matrix, which is formed on the basis on matrices of displacement, rotation, and scale.

For example, the loaded model increased in 1.05 times, then it moved along X-axis over two units and over one unit against Y-axis, and rotated by 10°, 6° and 10° around X-axis, against Y-axis and Z-axis respectively. The result matrix is:

$$STR_xR_yR_z = \begin{bmatrix} 1.03 & -0.2 & 0.07 & 0 \\ 0.18 & 1.02 & 0.2 & 0 \\ -0.11 & -0.18 & 1.03 & 0 \\ 2 & -1 & 0 & 1 \end{bmatrix}$$

Movement matrices formed when a user change the 3D model position. The generated matrices are transferred to workstation boards, where they applied to downloaded models.

7   Conclusion

In this article we have described the results of developing software and hardware part of our autostereoscopic display project, scientific and technical originality and stages of realization. It is described composite image forming and transformation using client-server technology. The proposed system implements three modes of interaction: static image demonstration, video playback and .3ds model loading and control. Based on the results of experiments it is proposed to create the workstation with single-board Raspberry computers. We have determined functions for the users computer and workstations, and have developed software for both of this systems.

It is assumed that the testing unit can be applied to determine the design decisions when creating the volumetric display, which may be used for visualization and control, decision support, based on the playback of volume, full-color, dynamic images in real time. A perspective is also the usage of this display in simulators, CAD and mechatronic systems.

Acknowledgment. This material is based upon work supported by The Skolkovo Innovation Center under Grant G65/15 of 3 September 2015.

References

1. Blundell, B.G., Schwarz, A.J.: Volumetric Three-Dimensional Display Systems. Wiley, New york (2013)

2. Bolshakov, A., Zhelezov, M., Lobanov, V., et al.: Development method of forming three-dimensional images for autostereoscopic volumetric displays. In: 2014 International Conference on Actual Problems of Electron Devices Engineering, vol. 2, pp. 461–468 (2014) (in Russian)
3. Bolshakov, A., Nikonov, A.: Three-dimensional display and method of forming three-dimensional images. Patent RF, N 2526901 (2013) (in Russian)
4. Nagano, K., Jones, A., Liu, J., Busch, J., Yu, X., Bolas, M., Debevec, P.: An autostereoscopic projector array optimized for 3D facial display. In: SIGGRAPH 2013 Emerging Technologies (2013)
5. Jones, A., Unger, J., Nagano, et al.: An automultiscopic projector array for interactive digital humans. In: SIGGRAPH 2015. ACM Press (2015)
6. Annen, T., Matusik, W., Pfister, H., Seidel, H.-P., Zwicker, M.: Distributed rendering for multiview parallax displays. In: Proceedings of SPIE 6055, Stereoscopic Displays and Virtual Reality Systems XIII (2006)
7. Lurie, A.I.: Analytical Mechanics. Springer, Berlin (2002)

YuMi, Come and Play with Me! A Collaborative Robot for Piecing Together a Tangram Puzzle

David Kirschner$^{(\boxtimes)}$, Rosemarie Velik, Saeed Yahyanejad,
Mathias Brandstötter, and Michael Hofbaur

Institute for Robotics and Mechatronics, Joanneum Research, Klagenfurt, Austria
{david.kirschner,rosemarie.velik,saeed.yahyanejad,mathias.brandstotter,
michael.hofbaur}@joanneum.at

Abstract. Human-robot collaboration is a novel hot topic in robotics research opening a broad range of new opportunities. However, the number of sensible and efficient use cases having been presented and analysed in literature is still very limited. In this technical article, we present and evaluate a collaborative use case for a gaming application in which a two-arm robot has to piece a Tangram puzzle together with a human partner. Algorithms and methods employed for this purpose are presented, performance rates are given for different setups, and remaining problems and future developments are outlined.

1 Introduction

Human-robot collaboration is a novel hot topic in robotics research opening a broad range of new opportunities. According to a market study of ABI Research, the "collaborative robotics" sector will increase roughly tenfold between 2015 and 2020. While envisioned future applications range far beyond today's main commercial focus of industrial production – addressing topics like assistive technology, service robotics, and consumer entertainment – current R&D still focus on equipping collaborative robots (so-called cobots) with relatively basic features, which are necessary for working safely in the proximity of humans (see Sect. 2). Furthermore, up to now, use cases demonstrating a sensible and efficient collaboration between humans and robots are still rarely presented and evaluated in literature.

In this article, we present and evaluate a collaborative use case for a gaming application. Concretely, a robot has to piece a Tangram puzzle together with a human partner. After the details of the addressed use case is introduced, the features and methods necessary for its implementation is outlined. In the following sections a discussion of achieved results and still existing limitations can be found. In the last chapter finally presents a conclusion and an outlook of further planned work.

2 The Emerging Field of Collaborative Robotics

An overview of the topic of "collaborative robotics" has to start with a discussion of the definition of this term. What exactly do people have in mind when talking

© Springer International Publishing Switzerland 2016
A. Ronzhin et al. (Eds.): ICR 2016, LNAI 9812, pp. 243–251, 2016.
DOI: 10.1007/978-3-319-43955-6_29

about a collaborative robot? According to the Oxford dictionary [3], the definition of the term "collaborative" (from Latin laborare: to work) is "produced by or involving two or more parties working together":

Until recently, this phrase was mainly used to describe the interaction and movement synchronization of multiple (static or mobile) robots assisting each other to perform a task that is either too difficult or impossible for one robot to perform alone (e.g. a particular welding procedure) [2,10]. Nowadays, the use of this term is changing and rather referring to a robot that operates with – or at least around – humans without fences or any additional safety devices. A number of commercially available robots declared as "collaborative" by their manufacturers, including their key features, are presented in [1]. Referring to human-robot collaboration, different levels of complexity can be distinguished for the interaction. Table 1 gives an overview of these levels as presented in [11]. Category A is in fact non-collaborative. The categories B to D address human-robot interaction with gradually increasing complexity. Category E describes the case of two collaborating robots (or one robot with two arms) and category F the interaction between two robots (or one robot with two arms) and one or more humans.

Table 1. Overview of interaction categories

Category	A	B	C	D	E	F
Umbrella term	Encapsulation	H-R co-existence	Static H-R collabo-ration	Dynamic H-R collabo-ration	Static/dynamic R-R collabo-ration	Static/dynamic H-R-R collaboration
Interaction level	Interaction-free operation	Safety stop	Static collabo-ration	Dynamic collaboration	Static/dynamic R-R colla-boration	Static/dynamic H-R-R collaboration
Actors	Robot	Human +robot	Human + robot	Human+robot	2 robots	2 robots +human(s)
Temporal depen-dence	Independent	Interrupt	Sequential	Simultaneous	Sequential/ simultaneous	Sequential/ simultaneous
Spatial depen-dence	Separated	Separated	Shared	Shared	Shared	Shared
Human-robot contact	None	Rudimentary	Pronounced	Comprehensive	N.a.	Pronounced/ comprehensive

Furthermore, according to the ISO standards [5,6], collaborative features are categorized into the following four basic types.

1. **Safety-rated monitored stop:** the safety-rated monitored stop feature is used to terminate the motion of the robot system when an operator enters the collaborative workspace.
2. **Hand guiding:** a hand-operated device is used to apply forces on the robot to guide or teach paths.
3. **Speed and separation monitoring:** allows the operator to work with the robot system simultaneously in the collaborative workspace.

4. **Power and force limiting:** This is a type of robot that is widely knows as collaborative robot. Physical contact either intended or incidental can occur between the worker and the robot system. Due to its built-in safety features, it can detect abnormal forces during operation.

However, only robots having implemented feature 4 are allowed to be operated without any additional safety devices. An example for such a robot is ABB YuMi®, which is used to manipulate the Tangram puzzle described in this work in collaboration with a human operator (see Sect. 4.1). In [4,8] more information can be found about this topic. Furthermore, in beginning of 2016, the new ISO standard 15066 [7] has been released, specifying parameters like maximum permissible force and pressure, with which an impact between human and robot may happen for not causing harm to the human body.

While robot manufacturers still focus mainly on equipping collaborative robots with the basic functionalities described above, a few research project (e.g. FourByThree[1], SecondHands[2]) are already on the way focusing on further, more advanced research questions of human-robot collaboration. One of these projects is the "Collaborative Robotics" [11] focusing on topics like (1) the dynamic perception of the environment, the human status, and the robot status, (2) an intuitive human robot interaction and information exchange, (3) dynamic adaptive planning, (4) sensitive redundant kinematic object manipulation. The use case and work presented in this article has been elaborated in the framework of this project.

3   Use Case Specification

The use case presented in this article is to solve a five-piece Tangram puzzle (Fig. 1(a)). For this purpose, a two-arm robot called ABB YuMi either solves this task alone (corresponding to the use case categories A, B and E from Table 1) or it collaborates with the human towards this goal (corresponding to the use case categories C, D and F from Table 1).

Initially, the puzzle pieces are placed by hand randomly in a circumvented area reachable by the robot and the human. At the beginning, the user can select the shape that shall be built with the puzzle (a square, a swan or any another predefined shape). The robot can also give feedback to the user (e.g. via the flexpendant, computer screen), for instance, in cases in which two asymmetric parts are placed in a way that no solution can be found or that not enough valid puzzle pieces are present in the game. It can also happen that more pieces are present in the game than necessary for building the puzzle. In this case, the robot and human select the pieces they need for completing the task. The building process stops when the human presses an emergency button or when the robot collides with the human or any other object. Figure 4 presents a formal representation of the different sequential steps necessary to realize the use case.

---

[1] http://fourbythree.eu/.
[2] https://secondhands.eu/.

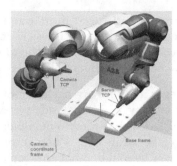

(a) Human-robot collaboration for building a Tangram puzzle

(b) Work objects in the workspace

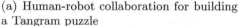

Fig. 1. ABB YuMi platform

4 Methods and Algorithms for Use Case Implementation

The choice of the robot manipulator and the vision system for puzzle piece recognition is crucial for the performance of the system. In this chapter, we give an overview about the used hardware, system setup, and parameter values set for a successful completion of the puzzle solving task.

4.1 Manipulation of the Puzzle Pieces

For implementing the specified use case, an ABB YuMi robot was used. Yumi is equipped with two lightweight and padded arms, which are controlled independently and are able to handle a payload of up to 500 g per arm. It has rounded and smooth surfaces to dissipate forces in case of an impact. The robot arms were equipped with two-finger grippers and a vision module integrated into the wrist of the right arm. Since the robot is designed to work in close contact with human, it is important that YuMi can pause its motion within milliseconds in case of collision detection. For this purpose, the robot measures the current of the motors in each joints during operation. In case of collision, a peak of current can be detected and the robot will stop instantly. The robot communicates with the human via a flex-pendant, indicating which part is currently searched for, reporting if some part is missing, or if the part having been detected has been removed in between. To program the robot, the official software from ABB (RobotStudio) was used. Offline programming gives the chance to simulate and test a program, predict collisions, or just simply define poses, while online programming allowed for debugging and tuning the actual robot motion.

4.2 Recognition of the Puzzle Pieces

To identify and locate the puzzle pieces, the camera embedded in the wrist of the right arm was used. The vision module is based on a Cognex In-Sight® ethernet smart camera. It has 1.3 Megapixel resolution, a 6.2 mm - f/5 lens, and integrated LED illumination. The camera has manual focus and therefore has to be calibrated for a certain position. In our case, it was calibrated for a distance of 500 mm from the table. This distance results in a field of view of approximately the size of an ISO A3 paper. In the integrated vision module of RobotStudio, programs loaded into the camera are called vision jobs. Each vision job consists of so-called location tools and inspection tools. Location tools provide position data of objects, like for instance, blobs, edges or patterns while inspection tools examine the located objects (measure distances, diameters or create geometric references, etc.). Every vision job has a limited size; it has to be taken into account that a more complex vision tool occupies more memory and requires more time to compute data. Before performing any vision job, a camera calibration is required. The calibration process consists of two steps: (i) geometric camera calibration to correct the lens distortion and (ii) calculating the extrinsic parameters of the camera to relate the camera coordinates to the robot frame. For our vision guided robotic tasks, determining the x-, y-coordinates and the orientation of the detected parts is of crucial importance. These three parameters define the object frame. In other words, after an image is captured and the required features are extracted, the object frame will provide the required data for the robot navigation toward the gripping position. These steps are shown in Fig. 2.

Fig. 2. Steps in the vision module

In RobotStudio, several "work objects" an be defined, which are in principle coordinate frames. When calling a move instruction (i.e. moving from point A to B), it is necessary to specify, which work object shall be taken as a reference. To simplify the program, we created a work object in the center of the camera image area, as shown in Fig. 1(b) and the positions for all vision jobs were calculated within the coordinate system of this work object (see Fig. 3(a)). To locate and inspect the puzzle pieces independently of angle, size, and shading, a pattern recognition algorithm based on "PatMax Pattern"[3] was used. This method learns an object's geometry based on a set of boundary curves, which are not tied to a pixel grid and are illumination-invariant. Afterwards, it looks for shape similarities in the image. After training the system on the occurring object shapes, the exposure time and the parameters of the vision tools such as accept threshold and offset values were tuned. Accept threshold defines degree of similarity between the taught and the found pattern. Selecting a higher threshold value leads to faster and more reliable results. However, a change in

---

[3] http://www.cognex.com/pattern-matching-technology.aspx?pageid=11368&langtype=2057.

lightning conditions may result in false negatives (no piece is detected while it is present). On the other hand, with a lower threshold value, the possibility of false positives increases. As shown in the example of Fig. 3(b), in case of a too low threshold and the absence of the triangle piece, the polygon is recognized as a triangle since the latter is a sub-shape of it. To overcome this error and ensure the individuality of the pieces, specific patterns on some parts were painted, see Fig. 3(a). Adding such patterns improved the robustness of detection under different lighting conditions.

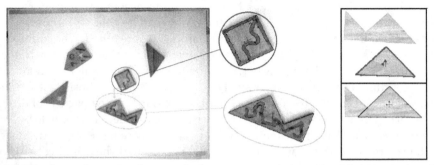

(a) The located puzzle pieces with the pattern and their object frames (b) Mismatch of the pieces

Fig. 3. Detected puzzle pieces

Another important parameter is the rotation tolerance, which was set to be rotation-invariant in the whole range of 360°. With horizontal and vertical offsets, the position of the object frame can be tuned. To simplify the program, we defined only one pick-up position for all parts. Since the robot moves in the object frame, we had to find an appropriate location of the frames for each part so that the robot is able to pick up the parts always in the same way. For some object shapes there may be only one solution for picking up some pieces with a two-finger gripper, such as the triangles in Fig. 3(a). After setting up the parameters, we taught the place-down position by "jogging" to the location where the puzzle should be solved. We set this location outside the camera detection area to avoid detecting the parts already handled. Note that computation time of a PatMax pattern is about 0.8 s and that not more than three PatMax patterns can be handled within one vision job due to time-out errors.

In the ABB integrated vision manual [9] two possible solutions to this problem are presented: (i) create an individual vision job for each part and then load it into the camera one after the other or (ii) use one vision job and disable the unused vision tools. In our experiments, we analyzed both solutions (see Table 2), but the second option seemed to be more time efficient. As described in the manual, the disabled tools still send information to the program, which should be sorted out by some measure. However, when implementing the reference solution presented in the manual, we could not achieve the desired result. Accordingly, we developed an alternative and additionally simplified solution to this problem: We consecutively called the "CamGetResult" command as many times as the sequential number of the disabled item in the list of outputs was. The list of outputs can be accessed in the vision module under the "Output to Rapid" command. Both solutions were tested with one arm as well as two arms and the results and timings were compared (see Sect. 5). For two-arm applications, synchronized movements are needed.

4.3   Collaborative Functions

Due to the hardware restriction, the robot is blind between two images, i.e., it may happen that the human pick up the same piece that the robot wants to. In such a case, the system can detect if it did not grab anything and check whether the piece exist on the target location or not.

On the other hand, the human can also assist the robot if an asymmetric piece is placed down with a wrong orientation or the forthcoming piece is not presented. We show that it is possible to fulfill these tasks with the mentioned hardware limitations and how the human-robot collaboration helps us to cope with these constraints.

5   Results and Discussion

In this section, first results concerning the developed robotic Tangram puzzle piecing algorithms are presented. In Table 2, four different algorithms are compared in terms of time required for successfully completing the puzzle piecing task. To make results reconstructible in these measurements, the same initial configuration of the puzzle pieces, the same lightning condition, and the same moving instructions with the same speed were used. Furthermore, the robot solved the puzzle without human intervention. As indicated in Fig. 4, one important prerequisite to achieve a robust recognition of puzzle pieces independent of the lightning condition was to set up the exposure time automatically in the beginning of the program by reading the brightness of the image. As expected, when only one robot arm was used to detect the pieces as well as to place them into the final position the robot was slower than when using one arm for image acquisition and the other one for puzzle placing. Furthermore, Table 2 illustrates the amount of time saving when using only one vision job for the puzzle part recognition instead of 5 vision jobs.

To analyse factors of human-robot collaboration in puzzle piecing, a first preliminary set of experiments has been performed with users, which shall at this point only be discussed briefly as far as relevant for the current algorithm. Further results will be presented in consecutive work.

During the experiments, the following shortcoming of the currently employed algorithms was identified, which should be subject to further improvement: During operation, the human operator is principally allowed to change the position of the pieces inside the image area and to take pieces away. However, these activities are only permitted within certain temporal boundaries. In case these adaptations occur between the time instant where the robot has taken an image of the current puzzle piece positions and the time instant where it picks up the next selected piece,

Table 2. Comparison of the algorithms

Method	Time
One hand - 5 vision jobs	78 s
Two hand - 5 vision jobs	64 s
One hand - 1 vision job	59 s
Two hand - 1 vision job	42 s

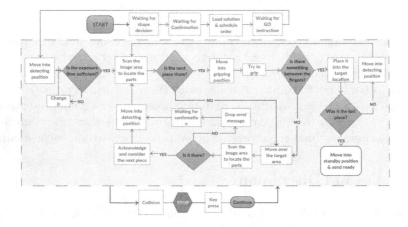

Fig. 4. The flowchart showing the steps and components for puzzle solving.

the robot can in the worst case collide with a piece at an unforeseen position and breaking its robotic finger.

6   Conclusion

In this article, a collaborative robotics use case for gaming applications was presented in which a two-arm ABB YuMi robot pieces a Tangram puzzle together with a human game partner. Employed methods for puzzle piece recognition, grasping, game sequencing, and emergency stop in case of collision were presented and evaluated. The next steps planned are amongst others the integration of a dynamic puzzle solver, natural human-robot communication concepts, and the integration of matrix-based proximity sensors for avoiding collisions between robotic and human game partners.

Acknowledgement. This work has been supported by the Austrian Ministry for Transport, Innovation and Technology (bmvit) within the project framework Collaborative Robotics.

References

1. Bélanger-Barrette, M.: Robotiq Collaborative Robot Ebook, 6th edn. Robotiq (2016). http://blog.robotiq.com/collaborative-robot-ebook
2. Cao, Y.U., Fukunaga, A.S., Kahng, A.: Cooperative mobile robotics: antecedents and directions. Auton. robots **4**(1), 7–27 (1997)
3. Oxford English Dictionary. Oxford University Press, Oxford (1989)
4. Fryman, J., Matthias, B.: Safety of industrial robots: from conventional to collaborative applications. In: Proceedings of the 7th German Conference on Robotics, ROBOTIK 2012, pp. 1–5. VDE (2012)
5. ISO 10218-1:2011: Robots and Robotic Devices–Safety Requirements Forindustrial Robots–Part 1: Robots. International Organization for Standardization, Geneva (2011)
6. ISO 10218-2:2011: Robots and Robotic Devices–safety Requirements Forindustrial Robots–Part 2: Robot Systems and Integration. International Organization for Standardization, Geneva (2011)
7. ISO 15066: Robots and Robotic Devices–collaborative Robots. International Organization for Standardization, Geneva (2016)
8. Michalos, G., Makris, S., Tsarouchi, P., Guasch, T., Kontovrakis, D., Chryssolouris, G.: Design considerations for safe human-robot collaborative workplaces. Procedia CIRP **37**, 248–253 (2015)
9. AAR Products: Application manual - integrated vision, Document ID: 3HAC044251-001, Revision: E (2015)
10. Ranky, P.G.: Collaborative, synchronous robots serving machines and cells. Ind. Robot Int. J. **30**(3), 213–217 (2003)
11. Rosemarie, V., Bernhard, D., Saeed, Y., Matthias, B., David, K., Lucas, P., Ferdinand, F., Patrick, L., Herwig, Z., Gerhard, P., Michael, H.: Step forward in human-robot collaboration the project collrob. In: OAGM & AWR Joint Workshop on Computer Vision and Robotics, Wels, Austria, pp. 1–8 (2016)

Author Index

Printed in the United States
by Publishers ...

Printed in the United States
By Bookmasters